THOMAS HENRY
HUXLEY

THOMAS HENRY HUXLEY

The Evolution of a Scientist

SHERRIE L. LYONS

59 John Glenn Drive
Amherst, New York 14228-2197

Published 1999 by Prometheus Books

Inquiries should be addressed to
Prometheus Books, 59 John Glenn Drive, Amherst, New York 14228–2197.
VOICE: 716–691–0133, ext. 207.
FAX: 716–564–2711.
WWW.PROMETHEUSBOOKS.COM

03 02 01 00 99   5 4 3 2 1

Library of Congress Cataloging-in-Publication Data

Lyons, Sherrie Lynne, 1947–
    Thomas Henry Huxley : the evolution of a scientist / Sherrie L. Lyons.
        p.   cm.
    Includes bibliographical references and index.
    ISBN 1–57392–706–6 (alk. paper)
    1. Huxley, Thomas Henry, 1825–1895. 2. Biologists—England
Biography. 3. Scientists—England Biography. I. Title.
QH31.H9L96   1999
570'.92—dc21   99–39254
[B]   CIP

Printed in the United States of America on acid-free paper

*In memory of my
mother and father,
Edna and Harold Lyons*

Acknowledgments

There have been many people who have helped me at various stages in the writing of this book. In its early form Jacob Gruber, Ernst Mayr, and Polly Winsor kindly answered my queries, and our correspondence raised issues I might not have considered. I benefited enormously from conversations with David Raup, David Jablonski, Russell Lande, Brian Charlesworth, William Provine, and most importantly Robert Richards. Will also loaned me several books which have been a great help. A wonderful group of graduate students in the geoscience department, the Morris Fishbein Center and Conceptual Foundations of Science at the University of Chicago provided an environment that was stimulating, challenging, and supportive. A special group of people read and reread numerous early versions, starting with a collection of disjointed essays. They are Douglas Allchin, John Brooks, Jeff Ramsey, and in particular Marc Swetlitz. If it had not been for Susan Abrams's early encouragement and support, a dissertation never would have become transformed into a book. David Hull greatly aided that transformation, and although I'm sure we will never completely agree on various issues concerning the type concept and systematics, his encouragement and critical reading of the manuscript have been invaluable. He forced me to deal with philosophical issues that as a historian I was all too willing to ignore. I have benefited from conversations and comments on specific chapters from Ron Amundson, Bernard Lightman, Michael Ruse, Phillip Sloan, and Peter Stevens. Peter Stevens sent me several wonderful articles relevant to classification. Charles Marshall has provided a continual update on developments in paleontology and clarified many issues for me. As the manuscript lay untouched for months collecting dust, correspondence with Jeremy Ahouse motivated me to find a publisher and finish it. The last chapter owes a particular debt to several people. Huxley had a more general, all-encompassing view of evolution than Darwin, and Ron Swenson suggested that I explore the implications of this for modern evolutionary theory, leading to the genesis of the final chapter. Dan McShea suggested several relevant readings, and he and Scott Gilbert critically read the final chapter, which resulted in important revisions. Robert Richardson read the entire manuscript and made useful suggestions. John Greene, in addition to providing many stimulating discussions, gave me a significant part of his own personal library. It would have been virtually impossible for me to have finished this book when I did if I had not had this most generous gift. As is usually said by authors (and I am no exception), the finished product is my own, and, therefore, whatever errors exist, I am responsible for them.

I want to thank the librarians at the American Philosophical Society, the University of Edinburgh, the British Museum of Natural History, and especially Anne Barrett at Imperial College of London. I always worked under severe time constraints, and they provided me with materials quickly and were extremely pleasant to work with.

In the time it took to write this book, my children, Cassandra and Grahame, have been transformed into young adults. Watching this transformation is the best example for me of why Huxley would be fascinated with the process of development. They are a continual source of joy and amazement and a daily reminder of what is truly important in this world.

Contents

Preface

Like any other discipline, the history of science exhibits trends. We have come a long way from the days of George Sarton, who decried the pseudosciences and believed that psychological, sociological, political, and economic factors were basically irrelevant to the writing of scientific history. Most current historians of science correctly deplore such a narrow approach to their craft. Such a limited approach is especially inappropriate regarding the work of a man such as Thomas Huxley. Huxley was an extremely complicated, charismatic man with far-ranging interests. Far more than Darwin, he was interested in the religious, philosophical, political, and socioeconomic questions of the day, and recent historiography has emphasized this aspect of Huxley's life. But Huxley was first and foremost a scientist. It was his science that brought him to the attention of people such as Charles Darwin, Charles Lyell, Richard Owen, and other members of the British scientific elite. Huxley was *the* premier advocate of science in the nineteenth century in the English-speaking world. That advocacy was the result of his stature as a scientist, even if it is also true that he used science to push his other agendas.

The production of scientific knowledge does not occur in a vacuum, and certainly to understand Huxley's ideas they must be situated within the larger framework of Victorian society. The controversies surrounding Darwinian evolution became a vehicle for Huxley to gain power in intellectual, institutional, and political arenas. However, we need to draw distinctions among why Huxley held the scientific beliefs he did, how he chose to defend those beliefs, and how those beliefs manifested themselves in the sociopolitical arena. Although these latter aspects of Huxley's career play a role in this study, the focus is on his scientific ideas. While current historiography has emphasized the sociopolitical context of Huxley's life, earlier scholarship lionized him as Darwin's advocate, portraying him as the heroic, idealistic defender of the "Truth." Both approaches have neglected Huxley's scientific corpus, which has not received the attention it so richly deserves. Thus, my goal is to provide an analysis of Huxley's research as Huxley's science is the foundation for the reputation of this most eminent Victorian.

*This paper would achieve one of the great ends of Zoology
and Anatomy, viz. the reduction of two or three apparent
widely separate and incongruous groups into modifications
of the single type, every step of the reasoning being based
upon anatomical facts.*
Thomas Huxley to his sister Lizzie, August 1, 1847[1]

Not Just Darwin's Bulldog

Self-appointed as Darwin's bulldog, Thomas Huxley claimed that "he was prepared to go to the stake if necessary" to defend Darwin's theory of evolution. For many, Huxley is known primarily as a popularizer of Darwin. However, he had a research program established long before he went to battle against the enemies of evolution. As the young assistant surgeon on the HMS *Rattlesnake*, he collected, dissected, and observed the development of a variety of invertebrates. But Huxley was no mere collector, writing up descriptions of his specimens in a series of technical monographs. As a letter to his sister Lizzie suggests and unpublished notes from the same period make clear, Huxley had an extremely ambitious agenda: he wanted nothing less than to provide a theoretical foundation for taxonomy. For Huxley, the most important task of zoological science was to define the doctrine of animal form. In order to do so Huxley's research placed him in the middle of one of the most important debates in nineteenth-century biology. Would a structuralist or functionalist approach yield the greatest insights in the quest to understand biological phenomena?

George Cuvier claimed that the "conditions of existence," by which he meant those conditions necessary for an organism to survive and reproduce, were responsible for functional adaptation, which in turn determined form. Morphologists, such as Karl Ernst von Baer, however, saw function as a byproduct of form. The commonalities of structure that existed among all living organisms and were referred to as "the unity of type" indicated a deeper reality in the understanding of biological form. A few basic types existed that limited the amount of functional adaptation possible. Although Darwin believed all life could be understood in terms of unity of type and the conditions of existence, he held the latter to be more important. Natural selection acted by adapting an organism to conditions of life, and the unity of type resulted from the inheritance of those adaptations. Thus, Darwin concluded that "the laws of the conditions of existence is the higher law."[2] Adaptation and natural selection has been Darwin's legacy. The success of natural selection theory in the twentieth century has resulted in the history of evolutionary theory being written from a neo-Darwinian point of view in which development plays no significant role. Because development would be the key to understanding biological form for Huxley, in such a history his research becomes hidden in the shadow of Darwin.

The nineteenth century was an extremely exciting time in biology. As Timothy Lenoir has written, the "focus on Darwin . . . has resulted in a distorted image of much important pre-Darwinian as well as a considerable body of significant post-Darwinian biology."[3] Spectacular discoveries in paleontology, anatomy, physiology, and embryology all contributed to Huxley's thinking about evolution. Rather than dismissing those of Huxley's views that were not in total agreement with Darwin's, a balanced assessment would place them within the context of the whole of nineteenth-century biology, not just Darwinism. Huxley's research was informed by a set of concerns that he shared with Cuvier and von Baer. The research of all three men suggested that organisms could be grouped into distinct types with no transitional organisms between them. Such findings led Huxley in his early years to argue against transmutation, the transformation of one species into another. However, unlike Cuvier and von Baer, who never accepted evolution, Huxley immediately embraced Darwin's bold new theory.

While Huxley enthusiastically accepted the idea of descent with modification, he was critical of the two central components of Darwin's theory: gradualism and natural selection. If organisms were grouped into distinct types, how could one organism gradually evolve into another? Huxley warned Darwin on the eve of the publication of the *Origin* that he had burdened his theory unnecessarily by adopting the dictum *Natura non-fecit saltum*, nature does not make jumps, and suggested that evolution by saltation (mutations or "jumps") better fit the facts of paleontology and embryology. Such a view also suggested that natural selection might be able to cause the formation of well-marked varieties but lacked the power to create true species. These doubts have led some recent scholars to claim that Huxley did not really understand the full implications of Darwin's ideas, and we find Huxley being labeled a "pseudo-Darwinian" or even "anti-Darwinian." I want to argue instead that Huxley's ideas reflect a tension between two different research traditions in biology that existed before Darwin, continue to exist today, and raise issues that remain problematic for evolutionary theory. Because of the success of natural selection theory in explaining adaptive complexity in the twentieth century, the functionalist approach has been

regarded as the more powerful. However, the idea of type provided a crucial organizing principle, led to the concept of
homology, and, as Darwin realized, could be used as evidence
for a theory of common descent. Huxley's work in morphology
provided evidence for Darwin's theory while reflecting
Huxley's own belief that a structuralist approach would yield
the greatest insights in the quest to understanding biological
phenomena.[4]

Although Huxley was more than Darwin's bulldog, nevertheless a large part of this book deals either directly or indirectly
with evolutionary theory. My reasons are twofold. First, in
some circles, Huxley's Darwinian credentials have been under
attack. Second, and more important, more than anyone else
Huxley was responsible for disseminating Darwin's theory of
evolution to the Western world. Indeed, Darwin early on
referred to him as his "general agent."[5] Huxley brought the
message of evolution not only to the learned societies, such as
the Royal Institute and the Geological and Zoological Societies
of London, but also to the larger public. He wrote numerous
articles on evolution for the popular press and lectured extensively before various improvement societies. His six lectures to
working men, "On Our Knowledge of the Causes of the Phenomena of Organic Nature" (1862), prompted Darwin to
write in dismay, "They are simply perfect. . . . What is the good
of my writing a thundering big book when everything is in this
green little book so despicable for its size? In the name of all
that is good and bad I may as well shut up shop altogether."[6]
Huxley toured America and gave a series of lectures on evolution that were printed in their entirety in the *New York Times*.
In the minds of scientists and the larger community Huxley's
name was linked to Darwin's. Nevertheless, Huxley *was* skeptical of the two basic tenets of Darwin's theory: natural selection and gradualism. Certainly this seemingly contradictory
stance needs explaining. Contrary to what much of the current
historiography on Huxley suggests, Huxley's own research was

deeply influenced by, and contributed to, evolutionary theory. His disagreements with Darwin over various aspects of his theory led not to a weakening, but rather to a strengthening of Darwin's thought. A good example is the dispute between Huxley and Darwin over the problem of hybrid sterility. For Huxley, true species were by definition incapable of inter- breeding or, if they could interbreed, produced offspring that were sterile. Huxley's continual badgering of Darwin over the problem of hybrid sterility led Darwin to conduct a series of experiments and to enlist the aid of other researchers to try and demonstrate that sterility was selected for which would aid process of speciation. (In other words, natural selection favored sterility in order to further the creation of diverse species.) Although Darwin returned to the initial view he presented in the *Origin*, that sterility was the byproduct of acquired traits, his position was far stronger because of these experiments.

Huxley's research is not just of historical interest. Many of the issues Huxley raised—such as what actually causes specia- tion, and whether the pattern of the fossil record is primarily saltational or gradual—continue to be actively debated among present-day evolutionary biologists. Finally, Huxley's first love—development—was not incorporated into the Modern Synthesis of Darwinian theory in the 1940s and remains one of the most difficult and interesting problems in biology. Exam- ining Huxley's work in the morphological tradition that makes use of the type concept can provide insight into current con- troversies over the relationship of developmental biology to evolution.

Organization of Chapters

This book is not meant to be a full-scale biography of Huxley but rather focuses on Huxley's scientific work. Huxley had interests that spanned virtually all areas of human knowledge.

During his lifetime, he played a crucial role in educational reform, particularly in advocating the teaching of laboratory sciences in the schools. He wrote and lectured on ethics, theology, and metaphysics, and he was involved in various political controversies. But fundamentally, Huxley was a scientist. The pursuit of scientific knowledge, he thought, gave meaning to life. He believed that it was only through the use of empirical techniques that we could secure for ourselves a little more ground of the previously unknown:

> All human inquiry must stop somewhere; all our knowledge and all our investigation cannot take us beyond the limits set by the finite and restricted character of our faculties, or destroy the endless unknown, which accompanies, like its shadow, the endless procession of phenomena. So far as I can venture to offer an opinion on such a matter, the purpose of our being in existence, the highest object that human beings can set before themselves is not the pursuit of any such chimera as the annihilation of the unknown; but it is simply the unwearied endeavour to remove its boundaries a little further from our little sphere of action.[7]

How did Huxley come to such a view? A brief biographical sketch in chapter 1 focuses on pivotal events in Huxley's life that shaped his character and contributed to his love of science. Particular attention is placed on Huxley's agnosticism. Huxley coined the word "agnosticism" to describe his own belief system, and it is his agnosticism that provides the framework for his scientific thought.

Chapter 2 situates Huxley's ideas within nineteenth-century biological thought by examining his early biological work, focusing on his belief in the concept of type. Cuvier's, von Baer's, and Huxley's own research were important influences in shaping his biological views and demonstrate that his belief in the morphological type concept was thoroughly grounded in the biological research of the early nineteenth century. Huxley's meaning of the word "archetype" is contrasted with that of the transcendental anatomists, in particular Richard Owen.

Until the publication of the *Origin* in 1859, Huxley argued against transmutation, in part because he believed organisms could be grouped into discrete types and that no transitional organisms existed between them. However, as Darwin recognized and Huxley came to see, the concept of type was perfectly compatible with evolutionary thinking. Pre-Darwinian and anti-Darwinian evolutionary theories have been described as developmental, characterized as being teleological and inherently progressive. However, the contrast between the developmental and Darwinian evolutionary views is not as clear-cut as some recent historians of nineteenth-century evolutionary biology suggest.[8]

Chapter 3 examines Huxley's belief in saltation. Saltation allowed Huxley to reconcile his belief in the concept of type with Charles Lyell's uniformitarian geology and Darwin's theory of transmutation. Lyell emerges as a crucial figure in Huxley's adoption of saltation. The historiography of nineteenth-century evolutionary biology has continually emphasized the gradualism of Lyell's uniformitarianism and its influence on Darwin. While Lyell believed geological changes were slow and gradual, he was a saltationalist in regard to the biological world.

One of the most heated scientific debates in the middle of the nineteenth century concerned the supposed progressive nature of the fossil record; in other words, the history of life documented a progression from simple organisms to more complex and higher ones. Many progressionists such as Adam Sedgwick, Richard Owen, and Louis Agassiz used the progressive nature of the fossil record as evidence for a revised version of the argument from design. The argument from design was a key aspect of the school of natural theology prominent in the 1700s. It had a significant resurgence with the publication of

William Paley's *Natural Theology* in 1802. According to this argument, each organism, and every part of each organism, had been designed for a specific adaptive purpose and was the product of divine wisdom. The order and complexity of the world, particularly as exemplified by living organisms, could not have come about by purely naturalistic means but was the result of the actions of an intelligent Designer. Chapter 4 analyzes Huxley's changing views on progression and reveals them to be the result of a complex interaction of a multitude of factors. Lyell's influence on Huxley extended beyond the issue of saltation. An ardent antiprogressionist, he convinced Huxley of the correctness of his views, and together the two men led the attack against progression in the 1850s. Both men marshaled impressive paleontological evidence in favor of their cause. However, as the century advanced, the case against progression became increasingly difficult to sustain, and Darwin's theory was a fresh assault on it. Huxley developed the idea of persistent types—organisms that had remained relatively unchanged for vast periods of time—to support nonprogression. Persistence also provided evidence for his belief in the type concept.

Both Huxley and Lyell denied progression partly because they wanted a totally naturalistic explanation of earth history. However, Huxley's reasons for his continued denial of progression were ultimately quite different from Lyell's. For the most part, Huxley's desire to draw distinct boundaries between science and theology served him quite well. But in chapter 4, I argue that his antitheological fervor got in the way of his objectivity, resulting in his maintaining an antiprogressive stance far longer than the evidence warranted.

Fossil evidence was crucial to Huxley's acceptance of progression. It also played a key role in his abandoning saltation. Chapter 5 examines Huxley's conversion to gradualism, discussing his work on *Archaeopteryx*, the dinosaurs, and horse phylogeny. Huxley's dinosaur research led him to propose that dinosaurs were the antecedent forms of reptiles and birds. The latter part of the chapter focuses on Huxley's famous American lectures on evolution, which, besides being brilliant expositions on evolutionary theory in all its nuances, illustrate that he used his popular addresses on Darwinism to deliver an even more fundamental message: Scientific questions can and should be kept distinct from theological ones.

Chapter 6 examines Huxley's role in the debates over human evolution and illustrates many of the themes that run throughout the book. First of all, Huxley was an empiricist. He believed that the classification of humans should be determined independent of any theoretical considerations; the theory of evolution or any other theory of human origins need not be invoked. Second, adopting such a position allowed him to remove the question of human ancestry from theological concerns. It was a strategy that he used again and again in his defense of evolution. His clash with Richard Owen played a prominent role in the debates over the ape-human question. While previous chapters touched on Huxley's disagreements with Owen, this chapter more fully discusses the relationship between the two men.

Huxley, of course, did more than just present "the facts." He argued passionately and eloquently that a common ancestry in no way degraded humankind. His lectures on man's place in nature superbly illustrate Huxley as a popularizer of science and in particular of evolutionary theory. In bringing Darwin's theory to the general public, Huxley influenced the scientific debates in ways that were different than if he had confined himself to more formal scientific writings. Scientific facts, personality conflicts, and philosophical beliefs were inextricably linked in the issues surrounding "man's place in nature."

Any treatment of Huxley's evolutionary views cannot ignore Huxley's doubts about natural selection. Chapter 7 examines the dispute Huxley had with Darwin over natural selection centering around the problem of hybrid sterility (i.e., sterile offspring of interbreeding species). As Mario di Gregorio and

Michael Ruse have previously claimed, Huxley's dispute with Darwin was primarily about what constituted proof of a hypothesis rather than overinterpretation of experimental results. Far from undermining his defense of evolutionary theory, Huxley's doubts about the efficacy of natural selection sharpened Darwin's own thinking about the problem of hybrid sterility. Huxley's skepticism resulted in Darwin performing a number of experiments that he might otherwise not have done, thus strengthening Darwin's own views. Although natural selection was not particularly relevant to Huxley's own research interests, his lack of interest did not weaken his support of Darwinism.

Chapter 8 returns to Huxley's philosophical and theological views, focusing on his agnosticism and his attitude toward materialism, showing how such beliefs shaped his defense of evolutionary theory. George Mivart raised many significant scientific objections to Darwin's theory, but Huxley's critique of Mivart largely ignored those objections, instead attacking Mivart's claim that evolution was compatible with the teachings of the Catholic church. Huxley's response to Mivart was exactly in the spirit of Mivart's attack on Darwin. Huxley's defense of Darwinism certainly promoted his own anticlericalism and his desire to remove theology from the practice of science. But it was also crucial that Darwin's theory be defended on just those grounds. Huxley recognized that many scientific objections were often at their root nothing more than a plea to put God back into nature. The practical consequence of Huxley's agnosticism was that he firmly believed the scientific method was the only useful way to increase one's understanding of the world.

In conclusion, chapter 9 examines the relevance of Huxley's work for modern biology by discussing some of the recent research in complexity theory and developmental genetics. Huxley's own view of evolution drew not just on the ideas of Darwin concerning biological evolution but also on a broader, more universal concept of evolution as espoused by Descartes and elaborated by Spencer. As Huxley recognized, developmental morphology was and continues to be crucial in studying problems in comparative anatomy, embryology, paleontology, and evolution. Recent research in complexity theory, the meaning of homology, and the resurgence of the idea of the morphogenetic field are attempting to fully incorporate embryology into modern evolutionary theory and in so doing are addressing precisely the sorts of issues that most interested Huxley.

Notes

1. *Life and Letters of T. H. Huxley* (*LLTHH*), 2 vols., Leonard Huxley, ed. (New York: D. Appleton & Co., 1901), 1: 36.

2. Charles Darwin, *The Origin of Species* (1859; New York: Avenel, 1976), p. 206.

3. Timothy Lenoir, *The Strategy of Life* (Dordrecht, The Netherlands: Reidel Publishing Co., 1982), p. 3. Peter Bowler also claims that Darwin's theory should not be considered the central theme in nineteenth-century evolutionary thought because natural selection was not widely accepted in Darwin's time. Bowler argues that while evolution was accepted, it was "within an essentially non-Darwinian conceptual framework . . . because it succeeded in preserving and modernizing the old teleological view of things" (*The Non-Darwinian Revolution* [Baltimore: Johns Hopkins University Press, 1988], p. 5).

4. See Ron Amundson, "Typology Reconsidered: Two Doctrines on the History of Evolutionary Biology," *Biology and Philosophy* 13 (1998): 153–77. Amundson argues that the recent historiography supports the position that most significant biological debate in Darwin's time was not evolution versus creation, but rather biological functionalism versus structuralism.

5. Charles Darwin to Thomas Huxley, March 4, 1860, Huxley Manuscripts 5.109, Imperial College of Science and Technology, London. The number "5.109" refers to William Dawson's catalog of the Huxley papers (London: Macmillan & Co. Ltd., 1946), and "HM" will be used in all future references to documents from the Huxley collection.

6. Charles Darwin, December 10, 1862, *LLTHH*, 1:223.

7. Thomas Huxley, *Six Lectures to Working Men* (London: Robert Hardwicke, 1863), p. 134.

8. See Bowler, *Non-Darwinian Revolution*, and Ernst Mayr, *The Growth of Biological Thought* (Cambridge, Mass.: Belknap Press, 1982). See Amundson, "Typology Reconsidered," for the contrary position, which is my own as well.

Chapter One

Rebel with a Cause

A Brief Biographical Sketch

T homas Henry Huxley was born on May 4, 1825, in the little country village of Ealing, about a dozen miles from Hyde Park Corner. He was the youngest of seven children of George and Rachel Withers Huxley. His father was the master at a large semipublic school with a excellent reputation. Physically and mentally, Thomas was his mother's child, having no trace of his father in him "except an inborn faculty for drawing, . . . a hot temper, and that amount of tenacity of purpose which unfriendly observers sometimes call obstinacy."[1] His mother, a slender brunette, was of an emotional and energetic temperament, which Thomas inherited. As was typical of middle-class women of her day, she was not especially well educated, but she had an excellent mental capacity. However, Huxley claimed that "her most distinguishing characteristic was rapidity of thought. . . . That peculiarity has been passed on to me in full strength."[2] Thomas also inherited his mother's wit, the trait he claimed to value the most.

In his autobiography, Huxley had "next to nothing to say about his childhood," which was for the most part uneventful.[3]

He attended his father's school for a few years before the family moved to Coventry. Most of his education was the result of his own private efforts. The Ealing School had fallen on hard times and eventually closed. Huxley had no kind words for it, claiming that the staff of the school "cared about as much for our intellectual and moral welfare as if they were baby farmers." The lads at school were rough: "We were left to the operation of the struggle for existence among ourselves, and bullying was the least of the ill practices current among us."[4]

In Coventry, Tom received little or no regular instruction, but he read everything he could get his hands on. At age twelve, he would rise before dawn, light a candle, and with a blanket around his shoulders sit up in bed and read James Hutton's geology.[5] From Sir William Hamilton's "The Philosophy of the Unconditioned" Huxley embraced the skepticism that typified his mature thought. From Thomas Carlyle he developed a sympathy for the plight of the poor and an intense hatred for shams of any kind. Carlyle also introduced him to German thinkers. He taught himself German in order to read people such as Goethe and Kant in the original. This would later serve him well, allowing him to be acquainted firsthand with the tremendous biological advances being made in Germany that few English men of science were able to follow. The young boy also had an intense interest in metaphysical speculation, and he discussed every imaginable question with family and friends. The nature of the soul, how it differed from matter, and the causes of colors in the sunset were all questions that occupied the young Huxley. His quick and eager mind enabled him to have friendships with those considerably older than himself.

Huxley was not going to school, but he was attending church. As he grew older, he realized with dismay that he might be one of those skeptics or infidels that preachers spoke of with such horror. Yet the preachers had left their mark. As William Irvine wrote, "In all but Doctrine Huxley remained a staunch Victorian Christian throughout his life. He was as morally earnest, as devoted to practical virtue, as suspicious of elaborate theological dogma as the most pious evangelical."[6] Huxley's desire to sermonize started at an early age. Young Tom fancied that he resembled the handsome courtly vicar of the local parish, Sir Herbert Oakley. "I remember turning my pinafore wrong side forwards in order to represent a surplice, and preaching to my mother's maids in the kitchen as nearly as possible in Sir Herbert's manner one Sunday morning when the rest of the family were at church."[7] Herbert Spencer later claimed that Huxley had strong clerical affinities, and there was some truth to the statement. Huxley preached his anticlericalism from every platform but the pulpit.

The young Huxley, like the mature one, had interests in a staggering array of subjects, but his deep curiosity about how things worked and how they acted resulted in his wanting to be a mechanical engineer. However, at a relatively young age he began to study medicine under the tutelage of a medical brother-in-law. Huxley showed a keen interest in human anatomy. Accompanying some older student friends, he attended a postmortem examination. In spite of being "unfortunately sensitive to the disagreeables which attend anatomical pursuits . . . my curiosity overpowered all other feelings, and I spent two or three hours in gratifying it."[8] Afterward, although he had not cut himself and had none of the ordinary symptoms of "dissection-poison," Huxley believed that he had somehow been poisoned and sunk into a strange state of apathy. It was so severe that he was sent to a farmhouse in Warwickshire to recover. He soon recuperated, but after that time suffered from "occasional paroxysms of internal pain" and "hypochondriacal dyspesia."[9]

In preparation for medical school, Huxley was apprenticed with Dr. Thomas Chandler, who had charge of the parish in the dock region of East London. Huxley was exposed to the squalor and poverty of life in the East End, and it deeply affected him for the rest of his life:

I saw strange things there—among the rest, people who came to me for medical aid, and who were really suffering from nothing but slow starvation. . . . Tall houses full of squalid drunken men and women, and the pavement strewed with still more squalid children. The place of air was taken by a steam of filthy exhalations; and the only relief to the general dull apathy was a roar of words—filthy and brutal beyond imagination. . . . All this almost within hearing of the traffic of the Strand, within easy reach of the wealth and plenty of the city.

I used to wonder sometimes why these people did not sally forth in mass and get a few hours' eating and drinking and plunder to their hearts' content before the police could stop and hang a few of them. But the poor wretches had not the heart even for that . . . drink and disease leave nothing in them.[10]

Experiencing life in the East End gave Huxley a "terrible foundation of real knowledge to [his] speculations" and committed him to a life-long advocacy of the poor and working classes.[11] He became a member of the London School Board and fought for education for the common man. He lectured extensively to the working classes in an effort to enrich their lives and improve their lot.

After two years of medical apprenticeship Huxley received a scholarship to the medical school attached to Charing Cross Hospital. His enthusiasm for anatomy did not carry over to the rest of the medical curriculum, however. The only part of his medical training that ever deeply interested him was physiology, which he described as the "mechanical engineering of living machines."[12] In his old age, he was "occasionally horrified to think how very little I ever knew or cared about medicine as the art of healing."[13] In spite of such a claim, Huxley did well at medical school. In 1843 he won the first chemical prize, as well as taking honors in anatomy and physiology. In 1845, he passed the first part of the Bachelor of Medicine exam at London University, winning a gold medal in anatomy and physiology. While at Charing Cross, he studied with Mr. Wharton Jones, a lecturer in physiology, the only instruction he ever thought valuable. The vast amount of Jones's knowledge and the exactness of his lectures were traits Huxley sought to emulate in his own lecturing. Jones was supportive of the young Huxley and suggested that he submit to the *Medical Times and Gazette* his first scientific paper, which described a layer in the root sheath of hair which has since been called Huxley's Layer. Huxley had been successful in all he undertook, yet he was still too young to qualify for a license to practice from the College of Surgeons!

Like many others who made their mark in the natural sciences, Huxley took a voyage around the world. He was deeply in debt, and an appointment at sea had its benefits. Through the influence of the distinguished naturalist Sir John Richardson, he enlisted in the navy and obtained the post of assistant surgeon on the HMS *Rattlesnake*, which set out on a surveying voyage in the south seas. The ship set sail in 1846 and did not return until 1850. During this time Huxley did some of his most important scientific work and met his future wife, Henrietta Anne Heathorn, while visiting Australia.

While on board the *Rattlesnake*, Huxley sent home various reports of his researches, two of which were published in the "Philosophical Transactions" of the Royal Society. Hence, by the time Huxley returned home, he had already established an excellent reputation. The following year, only twenty-five years old, he was elected a Fellow of the Royal Society and in 1852 was awarded one of the society's prestigious Royal medals. In spite of all these successes, this was not a happy time for Huxley. The navy, although agreeing to fund him to continue his research, refused to help finance the publication of his scientific treatises. Finally, in total frustration, Huxley deliberately refused to obey an order to board the ship *Illustrious*, effectively ending his relationship with the navy. Now that he had severed his relationship with the government, the Royal Society, which had previously been prevented from doing so by its regulations, could help finance his publications. Despite this support, Huxley had no permanent position and had no money to marry his beloved Netty. He was committed to following the path of a scientist and of settling in London. With only a handful of academic positions in biology even in existence, how long could he ask Netty to wait?

Fate intervened, though, and as Edward Forbes left the Royal School of Mines to take the chair of Natural History in Edinburgh, Huxley was appointed Professor of Natural History to the Royal School of Mines and Naturalist to the Geological Survey. He became Fullerian Professor of Physiology to the Royal Institution and a little later began to teach at the Royal College of Surgeons. His seven-year engagement to Netty finally came to an end with their marriage on July 21, 1855. When Darwin heard of Huxley's impending marriage, he said to him: "I hope your marriage will not make you idle; happiness is not, I fear, good for work."[14] Darwin's fears were not realized. Huxley was extraordinarily prolific in his career while enjoying an extremely happy home life.

From the time that Huxley returned from the *Rattlesnake* voyage to his retirement from various official posts thirty-four years later, he led a life of incessant activity. In the morning he lectured to students at the School of Mines. Between classes he continued to do research in physiology and biology. Evenings were often spent speaking before working men or learned societies. After he returned home he spent two or three hours preparing lectures or writing up the results of his research. Between 1863 and 1870, he was also a professor at the Royal College of Surgeons. In 1862, he was president of the biology section of the British Association, and in 1870 he became president of the association itself. In 1869 and 1870 he simultaneously was president of both the Geological Society and the Ethnological Society. In his lifetime he authored several hundred scientific monographs as well as countless popular essays.

Huxley's early work from the *Rattlesnake* voyage was

devoted to bringing some order to the chaos of invertebrate zoology. This work established his reputation within the scientific community, but it was his defense of Darwinism that brought him into the public spotlight. The famous encounter with Bishop Wilberforce at the 1860 Oxford meeting of the British Association for the Advancement of Science was an important milestone in Huxley's career. It was not just that he helped ensure that Darwin's theory received a fair hearing; "it was now that he first made himself known in popular estimation as a dangerous adversary in debate—a personal force in the world of science which could not be neglected. From this moment he entered the front fighting line in the most exposed quarter of the field."[15] After the publication of Darwin's *Origin of Species*, Huxley continued to publish monographs in virtually every area of zoology. While his early work focused mainly on invertebrate morphology, by the late 1850s he had started to work on fossils, and in the mid-1860s he published several important monographs on the dinosauria. Probably the most famous of Huxley's work is *Man's Place in Nature*. Published in 1863, eight years before Darwin published *Descent of Man*, *Man's Place in Nature* provided powerful evidence that humans were no exception to Darwin's theory.

However, Huxley's interests were not confined to science. He loved music, art, and literature. He especially admired Tennyson, claiming that he was the first poet since Lucretius "who has taken the trouble to understand the work and tendency of the men of science."[16] Huxley's house was a gathering place for the literary and artistic as well as the scientific elite of London. He also found time to be a member of two famous London clubs: the X Club and the Metaphysical Society. Ironically, the X Club was founded in 1864 at Huxley's suggestion to bring together a few scientific friends who were in danger of drifting apart because of their extremely busy schedules! These friends were the leading lights of a variety of scientific disciplines and included John Tyndall, Herbert Spencer, Sir John Lubbock, and Sir Joseph Hooker.

They dined together the first Thursday of each month except in the summer. The X Club soon developed the reputation for being "a sort of scientific caucus or ring." Huxley wrote,

> In fact two distinguished scientific colleagues of mine once carried on a conversation (which I gravely ignored) across me, in the smoking ring of the Athenaeum to this effect, "I say, A., do you know anything about the X club?" "Oh yes, B., I have heard of it. What do they do?" "Well, they govern scientific affairs, and really, on the whole, they don't do it badly." If my good friends could only have been present at a few of our meetings, they would have formed a much less exalted idea of us, and would I fear, have been much shocked at the sadly frivolous tone of our ordinary conversation.[17]

The club lasted until 1893. As members started to die and attempts were made to replace them, the remaining members realized that "the X really has no *raison d'être* beyond the personal attachment of its original members."[18] Huxley thought it would be best to let the club quietly die out unobserved. Joseph Hooker's comment best sums up the end of the X Club: "At our ages clubs are an anachronism."[19]

While the X Club members were all scientists, the Metaphysical Society drew its membership from all fields. Its members represented the intellectual elite of London society and included men such as William Gladstone, Alfred Tennyson, W. G. Ward, John Morley, and John Ruskin, as well as Huxley, John Tyndall, and John Lubbock. Victorians debated every possible question, but particularly the great question of science versus religion. Since the publication of the *Origin*, religion had been losing ground.[20] Huxley's "The Physical Basis of Life" had not helped the cause of religion either.[21] James Knowles, editor of the *Nineteenth Century*, suggested forming a society that would bring together the most distinguished men of the age genuinely concerned with religion. But the other side must be represented as well. "It was an appalling suggestion—a little like inviting the Devil to a debate on morality—but it was also in the most grandiose spirit of Victorian liberalism."[22] Thus, the Metaphysical Society was founded in 1869 and lasted eleven years. It was a perfect forum for Huxley to present his views on religion, the nature of knowledge, and the great theological and metaphysical questions that had occupied human thought since the beginning of recorded history. Huxley was an active participant, crossing swords many a time with Gladstone and Ward.

It would seem that there would be no time for family life, but somehow Huxley managed that as well. His wife was his confidante and critic whom he consulted on virtually every decision he made. His children were a joy to him. Congratulating Ernst Haeckel on the birth of his son, Huxley wrote: "Children work a greater metamorphosis in men than any other condition of life. They ripen one wonderfully and make life ten times better worth having than it was."[23] Huxley found family life a great source of pleasure and continually urged his bachelor friends to marry. After visiting the Huxley family in 1867 (which by this time included seven children, who ranged in age from ten years to infants), Dr. Anton Dohrn wrote:

> I have been reading several chapters of Mill's *Utilitarianism* today, and met with the word "happiness" more than once; if I had to give anybody a definition of this much debated word, I should say—go and see the Huxley family at Swanage; and if you would enjoy the same I enjoyed, you would feel what is happiness, and never more ask for a definition of this sentiment.[24]

Huxley's life was full of successes, but it was not without its difficulties or tragedies. In 1860, the year that brought him wide public recognition, he suffered perhaps his greatest sorrow. His son, not even four years old, died of scarlet fever. The suddenness of it, combined with his own anguish and his wife's inconsolable grief, brought him almost to a complete breakdown.

Three months later, the birth of another son gave Netty some comfort, and she conspired with John Tyndall for he and Huxley to go climbing in Wales for a week. Huxley's health was never good, and he suffered from frequent bouts of dyspepsia and depression. His frenzied life of appointments, courses, speaking, teaching, writing, and research finally took its toll, and in 1873 he suffered a complete breakdown.

In spite of all his appointments and lecturing, Huxley's finances were always somewhat precarious. Not only did he have his own large family to support, but he was regularly sending money to his sister Lizzie in the United States. In addition, he had a drunken sister to whom he sent money for most of her life and helped pay for her children's education. With all these strains on his purse, Huxley was too poor to go abroad for the rest that his physician Sir Andrew Clark had ordered. But the love and respect of his scientific colleagues came to his rescue. He received the following letter:

> My Dear Huxley—I have been asked by some of your friends (eighteen in number) to inform you that they have placed through Robarts, Lubbock & Co., the sum of £2,100 to your account at your bankers. We have done this to enable you to get such complete rest as you may require for the re-estab-lishment of your health; and in doing this we are convinced that we act for the public interest, as well as in accordance with our most earnest desires. . . . If you could have heard what was said, or could have read what was, as I believe our inmost thoughts, you would know that we all feel towards you, as we should to an honoured and much loved brother. I am sure that you will return this feeling, and will therefore be glad to give us the opportunity of aiding you in some degree, as this will be a happiness to us to the last day of our lives. Let me add that our plan occurred to several of your friends at nearly the same time and quite independently of one another.—My dear Huxley, your affectionate friend, Charles Darwin.[25]

Huxley, deeply touched and somewhat humbled, had to admit that anxiety and ill health had finally overcome him.

With botanist Joseph Hooker as nursemaid and a stack of medical instructions, Huxley set off to France. In spite of Huxley's effort to be cheerful, Hooker believed him to be severely depressed. But when Huxley found a copy of the *History of Miracles of Lourdes* in a Paris bookstall, he found something to engage his mind. "The prospect of a little congenial destruction both raised his spirits and quieted his digestion. Plunging happily into a great pile of treatises, he soon reduced all visions and cures to natural causes. Hooker perceived that his friend had healed himself by an act of unfaith."[26] The two men set off to the Auvergne and scaled extinct volcanoes, explored valleys for evidence of glaciation, smoked cigars, and generally had a splendid time. Traveling on to the Black Forest, Hooker finally left Huxley, who was then joined in Switzerland by his wife and son Leonard. After several months, his health was fully restored, and he returned to his whirlwind of activities. But Huxley would suffer another similar breakdown in 1884, and in 1885 he retired from government service.

Not long after this, another tragedy struck the Huxley family. Marian, Huxley's third child, a spunky and artistic girl who liked to cross swords with her father, had married painter John Collier in 1876. Not long into their marriage she suffered a complete psychotic break. For three years, her husband took her to all kinds of doctors and arranged for her to live in different places, all with the hope of her getting well. In the end Huxley even arranged for the famous physician Jean-Martin Charcot to treat her. Marian had gone to Paris where Charcot initially examined her and agreed to take her, but before she even went to his asylum, she caught a rapid pneumonia and died in 1887.

In 1889 Huxley built a villa at Eastbourne by the sea. He spent his last years gardening, with his grandchildren, and with his books. He continued to be intellectually active. Although ill health somewhat limited his activities toward the end of his life, Huxley completed editing a nine-volume set of his general works. Even during his final illness he was working on a criticism of Balfour's recently published "Foundations of Belief." In March 1895, Huxley came down with a case of bronchitis from which he was never able to recover. He died peacefully three months later, on June 30, 1895. Huxley's death brought tributes from around the world that revealed the multifaceted nature of the man. In his own time, Huxley's reputation was not just that of a great scientist or the defender of Darwinism, but that of a great humanitarian as well. Behind the sharp tongue was a man of tremendous kindness and charm. "Huxley to the last retained a charming simplicity of character that endeared him to all who were honoured by his friendship, and made him, even to the casual acquaintance, the most delightful of companions."[27] Huxley had often been a subject of the satiric writings appearing in the magazine *Punch*, but upon his death, *Punch* printed the following tribute:

> Another star of Science slips
> Into the shadow of eclipse!—
> Yet no; the *light* is nowise gone,
> But burning still, and travelling on
> The unborn future to illume,
> And dissipate a distant gloom,
> True man of Science he, yet more,
> Master of metaphysic lore,
> Lover of history and of art,
> He played a multifarious part,
> With clear head and incisive tongue
> Dowered, on all he touched he flung
> Those rarer charms of grace and wit
> Great learning may not always hit.
> To his "liege lady Science" true,
> He narrowed not a jealous view
> To her alone, but found all life

With charm and ethic interest rife,
Knowing plain lore of germ and plant,
With dreams of Hamilton and Kant,
All parts of the great human plan.
England in him has lost a Man.
The Great Agnostic, clear, brave, true,
Taught more things, may be,
       than he deemed he knew.[28]

Rebel with a Cause

[In the Metaphysical Society] every variety of philosophical
and theological opinion was represented . . . and expressed
itself with entire openness; most of my colleagues were *-ists* of
one sort or another; and however kind and friendly they might
be, I, the man without a rag of a label to cover himself with,
could not fail to have some of the uneasy feelings which must
have beset the historical fox when after leaving the trap in
which his tail remained, he presented himself to his normally
elongated companions. So I took thought and invented what
I conceived to be the appropriate title of "agnostic." It came
into my head as suggestively anti-thetic to the "gnostic" of
Church history, who professed to know so much about the
very things of which I was ignorant; and I took the earliest
opportunity of parading it at our Society, to show that I, too,
had a tail, like the other foxes. To my great satisfaction the
term took.[29]

Huxley's jocular account of how he coined the word "agnostic"
belies the seriousness of his desire to describe his own belief
system and to distinguish it from other -isms, such as positivism,
materialism, atheism and even empiricism. Typically, agnosticism
is placed on a religious spectrum somewhere between the absolute certainty of Christian belief and the total denial of the existence of God by the atheists. However, this was not the meaning

as Huxley originally intended. Rather, agnosticism represented an epistemological claim about the limits to knowledge. Building on the Kantian principle that the human mind had inherent limitations, and further elaborated by Hume, Huxley maintained that our knowledge of reality was restricted to the world of phenomena as revealed by experience.

Agnosticism is perceived as having an antireligious bias, in part because Huxley was well known for his polemics against theology. However, he took a perverse pleasure in pointing out that in coming to his conception of agnosticism he had been influenced by the High Church Anglican Henry Mansel's lectures on *The Limits of Religious Thought*, which emphasized what we don't and cannot know.[30] He agreed with Mansel, who argued that God's true nature was unknowable because as a transcendent being He was beyond the limits of human cognition. But God was not the only entity that was unknowable. As an epistemological claim, agnosticism did not apply just to Huxley's religious views; it provided the underlying framework for his life as a scientist. Agnosticism certainly challenged orthodox Christianity, but it also placed limits on the kinds of phenomena science could explain as well. For Huxley, atheists and materialists, just like theists and Christians, "were quite sure they had attained a certain 'gnosis,'—had more or less successfully solved the problem of existence; while I was quite sure I had not, and had a pretty strong conviction that the problem was insoluble."[31]

Theological and metaphysical questions interested Huxley, but he regarded science and philosophy as occupying distinct domains. As Jacob Gruber wrote, "Huxley made every effort to effect a complete separation between philosophy and science. For him . . . a sharp line marked the border between the provinces of empirical and metaphysical knowledge; and if the latter intruded upon the former both were corrupted."[32] On those questions that were not amenable to the scientific method, those that went beyond the cognizance of the five senses, he declared himself an agnostic. Although Huxley did not coin the word "agnostic" until 1869, journal entries and correspondence indicate that he had been thinking along such lines since the 1840s. In 1860 he articulated the basic principles of his agnosticism to Charles Kingsley in a letter about the immortality of man:

> I neither deny nor affirm the immortality of man. I see no reason for believing it, but, on the other hand, I have no means of disproving it.
>
> Pray understand that I have no *a priori* objections to the doctrine. . . . Give me such evidence as would justify me in believing anything else, and I will believe that. Why should I not? It is not half so wonderful as the conservation of force, or the indestructibility of matter. Whoso clearly appreciates all that is implied in the falling of a stone can have no difficulty about any doctrine simply on account of its marvellousness. . . . The universe is one and the same throughout; and if the condition of my success in unravelling some little difficulty of anatomy or physiology is that I shall rigorously refuse to put faith in that which does not rest on sufficient evidence, I cannot believe that the great mysteries of existence will be laid open to me on other terms. . . . Measured by this standard, what becomes of the doctrine of immortality?[33]

It was not that Huxley thought the question of immortality unimportant. Rather, he did not think it was fruitful to try and study problems that at the present stage of human knowledge were unsolvable. For Huxley, the existence of immortality was clearly such a problem. It could be argued, however, that for Huxley importance was synonymous with soluble. As Peter Medewar has written, science is the "art of the soluble." Huxley wrote Charles Kingsley, "You rest in your strong conviction of your personal existence, and in the instinct of the persistence of that existence."[34] But to Huxley such conviction meant nothing. He continued, "The attempt to conceive of what personality is, leads me into mere verbal subtleties. I have champed up all that chaff about the ego and the non-ego, about noumena and phenomena, and all the rest of it, too often not to know that in attempting even to think of these questions, the human intellect flounders at once out of its depth."[35] Thus, it was the intellect, not emotions or imagination, that would be useful in solving problems. But the intellect was "out of its depth" in trying to solve a problem such as the existence of immortality. Intelligent men could speculate endlessly on such questions with no resolution in sight. However, Huxley did not believe this was an antireligious stance. He went on:

> Science seems to me to teach in the highest and strongest manner the great truth which is embodied in the Christian conception of entire surrender to the will of God. Sit down before fact as a little child, be prepared to give up every pre-conceived notion, follow humbly wherever and to whatever abysses nature leads, or you shall learn nothing. I have only begun to learn content and peace of mind since I have resolved at all risks to do this.[36]

Science was a better instructor of the spirit than the Bible, church, or theology.

Huxley claimed that metaphysics and theology occupied a domain separate from science. He also drew a distinction between religion and theology.[37] He was first exposed to this distinction in Thomas Carlyle's *Sartor Resartus*, for him a crucial influence in his intellectual development. "*Sartor Resartus* led me to know that a deep sense of religion was compatible with the entire absence of theology."[38] Religion belonged to the realm of feeling while theology, like science, belonged to the realm of the intellect.

> All the subjects of our thoughts . . . may be classified under one of two heads—as either within the province of the intellect, something that can be put into propositions and affirmed or denied; or as within the province of feeling . . . called the aes-thetic side of our nature and which can neither be proved nor disproved, but only felt.[39]

Therefore, theology could conflict with science, but Huxley had no quarrel with religion "rightly understood," because it was not concerned with matters that could be subject to scientific investigation. Much to the amazement of many of Huxley's colleagues, he even advocated the reading of the Bible in elementary schools when he was a candidate for the London School Board in 1870:

> I have always been strongly in favor of secular education, in the sense of education without theology, but I must confess I have been no less seriously perplexed to know by what practical measures the religious feeling, which is the essential basis of conduct, was to be kept up, in the present utterly chaotic state of opinion on these matters, without the use of the Bible.[40]

Certainly pragmatic politics played a role in Huxley's advocacy of Bible-reading, and he later wrote that it was a "compromise" position. However, he made similar arguments in favor of reading the Bible in later essays and private letters. Such comments represented more than just a political strategy for getting elected to the school board. Huxley's heart and mind often were in conflict, and this is just one example where we find Huxley's emotional self doing battle with his intellectual self.

Huxley claimed the secularists were misguided in their demand for the abolition of religious teaching when they wanted only to free education from the Church and all it represented politically, socially, and intellectually. He did not advocate "burning your ship to get rid of the cockroaches."[41] In one of his many attacks on William Gladstone, Huxley even accused the opposition of "fabricating" the conflict between science and religion.

> The antagonism between science and religion, about which we hear so much, appears to me to be purely factitious—fab-

ricated, on the one hand, by short-sighted religious people who confound a certain branch of science, theology, with religion; and, on the other, by equally short-sighted scientific people who forget that science takes for its province only that which is susceptible of clear intellectual comprehension; and that, outside the boundaries of the province, they must be content with imagination, with hope, and with ignorance.[42]

Religion occupied a separate domain from science, while theology made empirical claims about the nature of the world, just as science did. Therefore, theological claims must be subject to the same standards of proof as scientific ones. Although Huxley many times articulated his distinction between religion (as feeling) and theology (as dogma), as author Ruth Barton correctly points out, he had far more interest in "attacking the dogma, rather than in developing the feeling."[43]

For Huxley, science was at war with theology, and there could be no compromise between the two. He associated theology with unquestioning following of authority, which in the practice of science was a totally unacceptable position. Belief in something because authority tells one it is true is a supreme virtue under the alias of "faith" but is totally inappropriate in the practice of the scientific method. Huxley paraphrased the bishop of Brechin's dictum " 'liberality in religion—I do not mean tender and generous allowances for the mistakes of others—is only unfaithfulness to truth' " by saying, "Ecclesiasticism in science is only unfaithfulness to truth."[44] In an anonymous editorial on "Science and Church Policy" in the *Reader*, Huxley wrote,

Science exhibits no immediate intention of signing a treaty of peace with her old opponent, nor of being content with anything short of absolute victory and uncontrolled domination over the whole realm of the intellect. Her champions ask why they should falter: Which of the memorable battles that have been fought have they lost?[45]

If some theological claim could be supported by "valid evidence and sound reasoning," then for Huxley it would take its place as a part of science. Theology would become "scientific" only when the Scriptures were treated as a collection of ordinary historical documents and analyzed using the research methods of philology, archaeology, and natural history. He could find support for such a view within certain theological circles. The higher criticism was suggesting that just such procedures be followed in interpreting the Bible. While Huxley applauded such developments in theology, he believed that the march of civilization was the result of the progress of science. Science represented the triumph of the natural over the supernatural, of fact over superstition, of knowledge over ignorance. And to borrow Ruth Barton's provocative title, Darwin's theory was "The Whitworth Gun in Huxley's War for the Liberation of Science from Theology."[46]

Huxley's antagonism toward natural theology was not a byproduct of his acceptance of Darwin's theory. Rather, his enthusiasm for Darwin was in part due to the absence of theology in Darwin's theory. As will be developed more fully in chapter 4, Huxley's desire for a totally naturalistic explanation of earth history in large part explains the virulence of his attack on *The Vestiges of Creation* in contrast to his laudatory review of the *Origin*. One reason that he continued to advocate the nonprogression of the fossil record, in spite of increasing evidence to the contrary, was because he linked progression with the argument from design.

Theological issues also played a role in Huxley's relationship with Richard Owen. Chapter 6 demonstrates that the hippocampus debate between Huxley and Owen was not merely about whether an ape had a hippocampus or not. Rather, Owen was attempting to use scientific arguments to imply a divine origin for humans. For committing such a sin, he received the full brunt of Huxley's wrath.

Although the findings in geology and paleontology were

certainly calling into question a literal interpretation of Genesis, the history of life could also be interpreted as powerful evidence for the argument from design. Moreover, theists claimed that their opponents had no adequate account of the origins of life. As Huxley wrote in "On the Reception of the 'Origin of Species,'"

> That which we were looking for, and could not find, was a hypothesis respecting the origin of known organic forms, which assumed the operation of no causes, but such as could be proved to be actually at work. We wanted, not to pin our faith to that or any other speculation, but to get hold of clear and definite conceptions which could be brought face to face with facts and have their validity tested.[47]

The *Origin* provided such an explanation.

> With any and every critical doubt which my skeptical ingenuity could suggest, the Darwinian hypothesis remained incomparably more probable than the creation hypothesis. . . . It was obvious that, hereafter, the probability would be immensely greater, that the links of natural causation were hidden from our purblind eyes, than that natural causation should be incompetent to produce all the phenomena of nature.[48]

Huxley enthusiastically championed Darwin's theory, in spite of having doubts about its most basic tenets, because it was "a powerful weapon which had destroyed a citadel of theology."[49] It presented a theory of descent in totally naturalistic terms. As I will discuss more fully in chapter 8, it is important to distinguish between the factors that shaped Huxley's scientific views on evolution and the factors that shaped his defense of Darwin's theory.

For Huxley, agnosticism was not a philosophy or creed. "In the sense of a philosophical system [it] is senseless: its import lies in being a confession of ignorance."[50] Such a statement explains, in part, why in spite of inventing the word, he was not all that involved with the many agnostic societies that sprung up. It was only at the end of his life that he wrote a series of essays that were specifically a defense of agnosticism.

While Huxley continually expressed that there were limits to knowledge, more often he emphasized that science would reveal the order in nature through empirical study of the physical world. Huxley claimed that science was not opposed to all theology, but he was opposed to what he called false theology. Huxley wanted a new theology based on the discoveries of science. Thus, he discussed the likelihood of miracles in relationship to the question of evidence and concluded that the evidence was wanting. His critical evaluation of evidence also made him skeptical of certain aspects of Darwin's theory in spite of accepting descent with modification. The next several chapters explore the nature of that evidence.

Notes

1. Thomas Huxley, "Autobiography," in *Collected Essays of Thomas Huxley*, vol. 1, *Methods and Results* (London: Macmillan & Co., 1889), p. 4.
2. Ibid.
3. Ibid.
4. Ibid., pp. 5–6.
5. Leonard Huxley, *Life and Letters of T. H. Huxley (LLTHH)*, 2 vols., Leonard Huxley, ed. (New York: D. Appleton & Co., 1901), 1: 6.
6. William Irvine, *Apes, Angels, and Victorians* (New York: Time Inc., 1963), p. 11.
7. Huxley, "Autobiography," p. 5.
8. Ibid., p. 7.
9. Ibid., p. 8.
10. Thomas Huxley, undated journal entry, *LLTHH* 1: 16–17.
11. Ibid., p. 16.
12. Huxley, "Autobiography," p. 7.
13. Ibid.
14. Charles Darwin, 1855 letter to Huxley, quoted in *Autobiography*

and Selected Essays by T. H. Huxley, S. E. Simons, ed. (New York: D. Appleton & Co., 1910), p. xiii.

15. Leonard Huxley, 1900, *LLTHH* 1: 193.

16. Thomas Huxley to Michael Foster, undated, but approximately October 1892 (after the death of Tennyson), *LLTHH* 2: 359.

17. Thomas Huxley, January 1894, "Reminiscences of Tyndall," quoted in *LLTHH* 1: 280.

18. Thomas Huxley to Joseph Hooker, 1888, *LLTHH* 1: 281.

19. Joseph Hooker, 1893, quoted in *LLTHH* 1: 282.

20. The reasons for the Victorian "crisis in faith" are complex and multifaceted, and many atheists and agnostics make little or no reference to science in explaining their loss of faith. A few relevant citings of the extensive literature that exists on the topic include John Hedley Brooke, *Science and Religion: Some Historical Perspectives* (New York: Cambridge University Press, 1991); Bernard Lightman, *The Origins of Agnosticism: Victorian Unbelief and the Limits of Knowledge* (Baltimore: Johns Hopkins University Press, 1987); Anthony Symondson, ed., *The Victorian Crisis in Faith: Six Lectures by Robert M. Young and Others* (London: Society for Promoting Christian Knowledge, 1970); and Frank Turner, *Between Science and Religion* (New Haven: Yale University Press, 1974).

21. I discuss "The Physical Basis of Life" in chapter 8.

22. Irvine, *Apes, Angels, and Victorians*, pp. 305–306.

23. Thomas Huxley, November 13, 1868, *LLTHH* 1: 32.

24. Anton Dohrn, 1867, *LLTHH* 1: 312.

25. Charles Darwin, April 23, 1873, *LLTHH* 1: 394–95.

26. Irvine, *Apes, Angels, and Victorians*, p. 342.

27. Obituary, *The Observer*, July 1895, Huxley Manuscripts (HM) 81:1:5.

28. *Punch*, July 13, 1895. HM 81:1:15.

29. Thomas Huxley, 1889, "Agnosticism," in *Collected Essays of Thomas Huxley*, vol. 5, *Science and the Christian Tradition* (New York: D. Appleton, 1898), p. 239.

30. See Lightman, *The Origins of Agnosticism*, for a thorough discussion of not just Huxley's views, but Victorian unbelief and the limits to knowledge.

31. Huxley, "Agnosticism," pp. 237–38.

32. Jacob Gruber, *A Conscience in Conflict: The Life of St. George Jackson Mivart* (New York: Columbia University Press, 1960), p. 21.

33. Thomas Huxley, September 23, 1860, *LLTHH* 1: 234.

34. Ibid.

35. Ibid.

36. Ibid., p. 235.

37. For an excellent discussion on the distinction Huxley draws between religion and theology see Ruth Barton, "Evolution: The Whitworth Gun in Huxley's War for the Liberation of Science from Theology," in *The Wider Domain of Evolutionary Thought*, D. Oldroyd and I. Langham, eds. (Dordrecht, The Netherlands: D. Reidel Publishing Co., 1983), pp. 261–87.

38. Thomas Huxley to Charles Kingley, September 1869, *LLTHH* 1: 237.

39. Thomas Huxley, 1882, "On Science and Art in Relation to Education," in *Collected Essays of Thomas Huxley*, vol. 3, *Science and Education* (London: Macmillan & Co., 1905), p. 175.

40. Thomas Huxley, 1870, "The School Boards: What They Can Do, and What They May Do," in *Critiques and Addresses* (New York: D. Appleton & Co., 1873), p. 51.

41. Ibid., p. 49.

42. Thomas Huxley, 1885, "Mr. Gladstone and Genesis," in *Collected Essays of Thomas Huxley*, vol. 4, *Science and the Hebrew Tradition* (New York: D. Appleton & Co., 1897), pp. 160–61.

43. Barton, "Evolution," p. 265.

44. Thomas Huxley, 1871, "Mr. Darwin's Critics," in *Collected Essays of Thomas Huxley*, vol. 2, *Darwiniana* (London: Macmillan & Co., 1893), p. 149, quoting Bishop Brechin's Charge at the Diocesan Synod of Brechin.

45. Quoted in Barton, "Evolution," p. 267.

46. The Whitworth gun was a rifle used by the British in the 1860s with a new muzzle-loading design that resulted in greater accuracy (ibid., fn. 1, p. 281).

47. Thomas Huxley, "On the Reception of the 'Origin of Species,' " in *Life and Letters of Charles Darwin (LLCD)*, 2 vols., Francis Darwin, ed. (New York: D. Appleton & Co., 1990), 1: 550–51.

48. Ibid., p. 551.

49. Barton, "Evolution," p. 281.

50. HM 30: 152–53, quoted in Lightman, *Origins of Agnosticism*, p. 13.

Chapter Two

Huxley and the Concept of Type in the Nineteenth Century

O n the eve of the publication of *The Origin of Species*, Huxley cautioned his good friend Charles Darwin, "You have loaded yourself with an unnecessary difficulty in adopting *Natura non facit saltum* so unreservedly."[1] In the years immediately following the publication of the *Origin*, Huxley praised Darwin's theory but voiced doubts about certain points in both popular and technical discussions. "Nature does make jumps now and then and a recognition of that fact is of no small import in disposing of many minor objections to the doctrine of transmutation," Huxley wrote in his famous *London Times* review of the *Origin*.[2] His caution to Darwin is often quoted, but without a critical analysis of the reasons behind the warning. The following two chapters provide such an analysis, not in relation to Darwin's gradualism, but rather in light of the other influences that shaped Huxley's thought. Saltation allowed Huxley to reconcile his belief in the existence of distinct morphological types with Darwin's theory of evolution. Furthermore, I will argue that it was Charles Lyell, whose name is usually associated with gradualism, who urged Huxley to his saltational view.

Huxley, like many others associated with Darwin, often stands in Darwin's shadow. But, as I mentioned in the introduction, Huxley was not merely Darwin's bulldog. Long before the publication of the *Origin*, Huxley was immersed in his own studies of developmental morphology and had been heavily influenced by Georges Cuvier and Karl Ernst von Baer. In championing Darwin, Huxley described Darwin's work as "the greatest contribution which has been made to biological science since the publication of the *Règne Animal* of Cuvier, and since that of the *History of Development* of von Baer."[3] Toward the end of his life, when asked to evaluate Darwin's position in the history of science, Huxley thought von Baer "would run [Darwin] hard . . . in both breadth of view and genius" and ranked Cuvier only slightly lower than Darwin.[4]

Judging Huxley only in relation to Darwin does an injustice not only to Huxley, but also to the history of biological science. In particular, as Lenoir has written, "the emphasis on Darwin is in large part responsible for the fact that we have overlooked a significant, valid alternative approach to biological phenomena during the early nineteenth century."[5] In early-nineteenth-century biology, Lenoir argues, two approaches were lumped together under the terms "transcendental morphology" and "idealistic morphology": that of the *Naturphilosophen* and what he called the teleomechanical approach of Kant and Blumenbach. Although his distinction has been criticized, in part because both traditions were teleological, Lenoir's analysis can be useful in evaluating Huxley's place in nineteenth-century biology—particularly in understanding how Huxley interpreted von Baer's work.[6] In any case, Lenoir's work provides a well-needed corrective to the overemphasis that has been placed on selection. The importance of selection was an open question, while research in developmental morphology was and continues to be crucial in studying problems in comparative anatomy, embryology, paleontology, and evolution.

Like Richard Owen, Louis Agassiz, and von Baer, Huxley was trained in the tradition of idealistic morphology. He was primarily influenced by the work of von Baer and physiologist

Johannes Müller. Müller had also been indoctrinated with the speculative doctrines of the *Naturphilosophen*, but his views on that approach to nature had been modified by the time he received the *venia legendi* in 1824.[7] Müller believed biology should be an observational and experimental science—a claim Huxley also made throughout his career. Owen also drew from von Baer and Müller, but he was sympathetic to the theoretical claims of *Naturphilosophen* made by J. W. Goethe, Frederick Tiedemann, and Lorenz Oken, claims and claimants Huxley had no use for. Cuvier, while not of the same tradition as these other men, shared with them a search for unifying laws to explain the organic world, an approach that was attractive to the empirically minded young Huxley. The idea of the organic type captured the imaginations of very diverse thinkers, including Huxley. The following sections examine the development of the type concept in the nineteenth century and how Huxley applied it to his own work.

The Concept of Type

The concept of type has received considerably bad press, largely due to the writings of the biologist and historian Ernst Mayr. According to Mayr, Darwin replaced typological thinking with population thinking, and in so doing he "produced one of the most fundamental revolutions in biological thinking."[8] Variation, writes Mayr, is "irrelevant and uninteresting" to the essentialist.[9] But in making such a claim Mayr blurs the distinction between the different meanings of type.[10]

There were at least three distinct usages of the type concept in the early part of the nineteenth century. The first was the classification type concept. This was a form that could be used as a model for a particular group and could be applied to several taxonomic levels. William Whewell in the *Philosophy of the Inductive Sciences* defined type as "an example of any class, for instance, a species of a genus, which is considered as eminently possessing the characters of the class."[11] The classification type concept helped naturalists organize their material rationally and also doubled as a name carrier in nomenclature. A second use of type, the type specimen, was developed as a result of the enormous increase in the size of museum collections and the need for an accurate identification of a particular specimen to name new species. The type specimen or collection type concept was similar to the classification concept, but it referred to a particular specimen in a known collection rather than to a general group. The type specimen was restricted to the species level and was narrowly defined.[12]

The third usage of type, however—the morphological type concept—is what interested Huxley. Comparative anatomists were interested in finding a basic plan or type that was discernible at different taxonomic levels and that could be used to explain the overall organization and functioning of an organism. As Paul Farber wrote, the morphological type concept was useful because it

> was an organizing principle for comparative anatomy. For some it held out the hope of raising comparative anatomy beyond a descriptive stage to an understanding of nature's laws regulating form. For others it represented a deep insight into nature's laws and suggested a key for unifying physiology, anatomy, and classification, and perhaps distribution and paleontology as well.[13]

Huxley, in his early years, like other pre-Darwinian morphologists, was searching for a classificatory scheme that reflected the natural groupings of organisms based on the similarity of structure. He believed such a system of natural classification was possible precisely because organisms could be grouped into distinct types.

Most of the abuse that has been heaped on the type concept

has been directed at a usage that traces its roots back to Plato. According to Plato, natural phenomena were reflections of fixed and unchanging forms or *eidoi*. The essence of an organism was an independent ideal form or type; living organisms represented variants and departures from the underlying essence. This concept of type was ubiquitous in the nineteenth century, with the *Naturphilosophen* being its most conspicuous advocates. On the Continent, Cuvier, Goethe, Oken, von Baer, Agassiz—everyone—was promulgating this concept of type. In Great Britain, Owen promoted the ideas of the *Naturphilosophen*.[14] However, Huxley was "unable to see the propriety and advantage of introducing into science any ideal conception which is other than the simplest possible generalized expression of observed facts." He viewed with extreme aversion "any attempt to introduce the phraseology and mode of thought of an obsolete and scholastic realism in biology."[15] Huxley had no use for the metaphysical claims of the *Naturphilosophen*. The notion of Platonic archetypes, he asserted, "is fundamentally opposed to the spirit of Modern Science."[16]

While von Baer must certainly be considered a member of the *Naturphilosophen*, his work, like Cuvier's, used a concept of morphological type that Huxley would adopt in his own work. When Huxley referred to archetypes in his own work, it was clearly von Baer's usage he had in mind. Morphology, Huxley wrote, "demonstrates that the innumerable varieties of the forms of living beings are modeled upon a very small number of common plans or types ('hauptypen' of von Baer, whose idea and terms are merely paraphrased by 'archetype,' common plan, etc.)."[17] "The doctrine that every natural group is organized after a definite archetype," he added, is "a doctrine which seems to me as important for zoology as the theory of definite proportions for chemistry."[18] However, Huxley wanted to distinguish his type concept from that of the transcendental morphologists, disclaiming any reference to "real or imaginary ideas upon which animal forms were modeled."[19] Rather, he thought of the type as a diagrammatic representation of the general characteristics of a particular group such as the mollusks. The archetype had the same relationship to the actual class of organisms "as the diagram to a geometrical theorem, and like it, at once imaginary and true."[20] Describing the general anatomical uniformities of a group of worms, the ascidians, Huxley said they were capable of being represented by a diagram. This hypothetical structure he called the "archetypal ascidian," and from it every actual form could be shown to be derived by simple laws of modification. To be sure that his archetype was not confused with that of the idealists, he "particularly desired it be understood that he attached no other meaning to the term archetype than that thus defined."[21] However, Huxley's views were not as different from the idealists as he would have us believe. Goethe's concept of type was not a metaphysical entity but a scientific artifice. Transcendental anatomist Carl Carus also described his vertebrate archetype as the "simplest schema."[22]

If many of the transcendentalists claimed that their type concept was nothing more than a diagrammatic representation of the general characteristics of a particular group, why was Huxley so anxious to distance himself from them? For many of the idealists the concept of type was evidence of "nature's divine plan," and Huxley did not want his archetype used as support for the argument from design. In particular, he singled out Richard Owen when he attacked the ideal notion of type. In practical terms, however, in his attempt to classify organisms and discover the underlying principles of animal organization, his use of the morphological type concept differed little from Owen's. I shall return to this point later.

The type concept was not merely a tool for the purposes of classification. In the 1840s and 1850s Huxley followed Cuvier and von Baer in arguing not only that distinct types existed, but also that they were fixed. Although von Baer allowed for a limited amount of evolution within groups, there could be no transformation of one group to another. Cuvier and von Baer did not argue for the fixity of type from idealistic principles or religious grounds. Rather, they made their case from empirical evidence. It may be true that they had enough theoretical assumptions to determine what empirical evidence would count. In any case, the young Huxley found that evidence quite convincing.

The Classification of Cuvier

E. S. Russell, in his history of animal morphology, described Cuvier as "perhaps the greatest of the comparative anatomists."[23] Cuvier's monumental *Le Règne Animal* (1817) reorganized the classification of the entire animal kingdom. It quickly became the standard manual of zoological systematics and was used with some modifications for most of the nineteenth century.[24] Huxley had carefully studied Cuvier and referred to him as the "Prince of modern naturalists."[25]

For Cuvier, functional necessity determined the structure of an organism, and any unity of plan or symmetry that existed within *embranchements* or types was a result of the "conditions of existence"[26]—those conditions necessary for an organism to survive and reproduce. Cuvier admitted that this was really just another term for final causes. The Creator had created just those organs that were needed for an organism to survive in any particular environment. There was no place for useless organs in Cuvier's thinking.

Cuvier classified the animal kingdom into four *embranchements* or types: vertebrata, mollusca, articulata, and radiata. He described the type genus by describing its general characteristics. A family was an array of similar genera, and the type species was the most representative species of each genus.[27] The divisions within the *embranchement* were "rather slight modifications founded on the development of the addition of some parts, but which change nothing of the essence of the plan."[28]

Cuvier did not deny individual variation within a species, but he regarded such variations as restricted by the type of animal it was. In his view individuals represented the species, and species represented the *embranchements*. For Cuvier, the *embranchement* was not a taxonomic category that had been invented for the purposes of classification. The *embranchements* were just as "real" as species.[29] Therefore, the fixity of the species was determined by the *embranchements*. For Cuvier, the *embranchement* was "what brought nature together."[30] But since they were fixed there could be no transmutation from one to the other.

Cuvier's *embranchements* were not the abstract types of the pure morphologists. He did not even use the word "type" in his work.[31] Organisms shared basic plans only because they carried out similar interrelated functions. Cuvier based his system on a hierarchy of anatomical characteristics, with the nervous system being the most important. He claimed to have found four different types of nervous systems, hence his division of the animal world into four *embranchements*. Within each *embranchement* subordinate morphological types could be determined by using other organ systems.[32] In Cuvier's scheme, functional adaptation was what determined the structure of the organism. Cuvier's famous principle of the correlation of parts reflected his belief that every organ in the body was related and dependent on every other organ in order to maintain the functional integrity of the organism as a whole. It was not possible to modify one organ without modifying others and still maintain harmony. This principle explained the gaps found between various groups, particularly between the *embranchements*. "There is no nuance between the second *embranchement* and the first nor any resemblance in the general disposition of parts."[33] Cuvier's scheme sought to explain the forms of animals, and in doing so it represented a *natural* system of classification.

Although impressed with Cuvier's work, Huxley asserted that Cuvier had arrived at his "great results" by a different method than what Cuvier had claimed.[34] Cuvier claimed functional adaptation underlay his classification and that his restorations of extinct animals were based on his principle of the physiological correlation of organs. However, Huxley argued that the idea of morphological constancy, rather than functional adaptation, had guided Cuvier in his classification. Huxley quoted Cuvier's own arguments and then proceeded to analyze them:

> It is readily intelligible that ungulate animals must all be herbivorous, since they possess no means of seizing a prey. We see very easily also, that the only use of their fore feet being to support their bodies, they have no need of so strongly formed a shoulder; whence follows the absence of clavicles. . . . No longer having any need to turn their fore-arm, the radius will be united with the ulna, or least articulated by a ginglymus. . . . Their herbivorous diet will require teeth, with flat crowns, to bruise up the grain and herbage; these crowns must needs be unequal and to this end enamel must alternate with bony matter; such a kind of crown requiring horizontal movements for trituation, the condyle of the jaw must not form so close a hinge as in the carnivore; it must be flattened and this entails a correspondingly flattened temporal facet.[35]

But Huxley maintained that Cuvier's conclusions were not based on physiology at all. Huxley asked why ungulates could not be carrion feeders. And even if they preyed on live animals, "surely a horse could run down and destroy other animals with at least as much ease as a wolf."[36] Other than support, what was the purpose of the forelegs of the dog and wolf, and how large were their clavicles? Huxley pointed out that the sloth was an herbivore, but its teeth did not show any traces of the alterations that Cuvier described. Furthermore, there was no difference in the "structure of tooth, in the shape of the condyle of the jaw, and in . . . the temporal fossa, between the herbivorous and carnivorous bears."[37] Huxley maintained that if bears were only known as fossils, no anatomist would conclude from the skull and teeth alone that the white bear was naturally carnivorous, while the brown bear was frugivorous (i.e., herbivorous).

Huxley's comments raised a fundamental question of taxonomy: What should be primary—morphological constancy or functional importance? Huxley may not have disputed Cuvier's theoretical claim that function determined structure. But he did argue that there was no way to empirically test the validity of such a claim. Huxley realized that although Cuvier claimed that function was primary in his classificatory scheme, it appeared that it was actually based on morphological constancy. First of all, the physiological criterion, founded on the correlation and subordination of parts, often failed. Besides, as Huxley had shown with his example, there was no way to actually demonstrate the physiological function of particular organs in extinct animals. Rather, countless observations of all variety of organisms indicated that certain structures were virtually always found in association with one another. As he pointed out,

> Our method then is not the method of adaptation, of necessary physiological correlations; for of such necessities, . . . we know nothing; but it is the method of agreement; that method by which having observed facts invariably occur together, we conclude they invariably have done so, and invariably will do so.[38]

Support of Huxley's claim regarding Cuvier was later brought by historian of comparative anatomy Henri Dauden.[39]

Although Huxley disputed Cuvier's concept of utilitarian adaptations and what Cuvier claimed was his method of classification, it must be emphasized that Huxley did not dispute Cuvier's basic conclusions. Huxley was impressed with Cuvier's attempts to order the natural world. He wrote Darwin that "Cuvier's definition of the object of classification seems to me to embody all that is really wanted in science—It is to throw the facts of structure into the fewest possible *general propositions*."[40]

Whether Cuvier determined his classification by means of physiological correlation of parts or by comparative morphology as
Huxley suggested, the end result was not disputed. Regardless
of how he achieved his restorations, Cuvier's fossil work indicated that large gaps in the fossil record existed with little evidence of transitional organisms. Indeed, this was a major
problem for Darwin. As he admitted in the *Origin*: "Geology
assuredly does not reveal any such finely graduated organic
chain; and this perhaps is the most obvious and gravest objection which can be urged against my theory."[41] Darwin devoted
two chapters in the *Origin* to show that while the fossil record
might not be cited in support of his theory, it could at least be
argued that it was not against it.

Cuvier conceptualized types as stable patterns of organization. Working within such a framework, it was difficult to
imagine one type changing into another. Like Cuvier, Huxley
believed there could be modification within a type, but no transition between types. "As Cuvier long ago remarked of the
Cephalopoda and Fishes so we may say of the Cephalous mollusca in general and other types. 'Whatever Bonnet and his followers may say Nature here leaves a manifest hiatus among her
productions.' "[42] In a footnote Huxley left no doubt as to his
beliefs concerning transitional forms:

> It is one thing to believe that certain natural groups have one
> definite archetype or primitive form upon which they are also
> modeled; another, to imagine that there exist any transitional
> forms between them. Everyone knows that Birds and Fishes
> are modifications of the one vertebrate archetype; no one
> believes that there are any transitional forms between Birds
> and Fishes. I beg that I may not be misunderstood here.
> While I consider that there is no transition between the
> Cephalous Mollusca as such, and the Ascidians or Polyzoa, I
> also fully believe . . . that the archetype of the Cephalous
> Mollusca, that of the Ascidians and that of Polyzoa, are all

referable to a common archetype, the archetype of the Mollusca generally.[43]

Although he would later change his views, at this point in time Huxley believed that his own work, like Cuvier's, demonstrated not only the reality of type; it was strong evidence against transmutation as well.

Von Baer and Development

While Cuvier's work supported Huxley's belief in type, Huxley's own usage of the type concept followed most closely that of von Baer. Extremely impressed with von Baer's work, Huxley in 1853 translated von Baer's Fifth Scholium and part of the Sixth Scholium of the *Entwickelungsgeschichte der Thiere*.[44] Like Cuvier, von Baer's research also indicated there were four basic animal plans. But the two differed in certain important respects. Cuvier's *embranchements* were based on the constant association of interdependent anatomical structures and the interdependent functions revealed by comparative anatomy and physiology of adult organisms. But von Baer discovered his four basic types in developing embryos. The type was

> the relative position of the organic elements and of the organs.
> . . . The type is totally different from the grade of development
> so that the same type may exist in many grades of development. *The product of the grade of development with the type yields those separate larger groups of animals which have been called classes* (emphasis in original).[45]

Classes of organisms thus represented different grades of development within types. Von Baer generally regarded the vertebrate type as being "above" the invertebrate type. He recognized, however, that the muscles and nerves were more differentiated in bees than in fish, hence bees represented a higher grade of development in the articulate type.[46] Four archetypes, he wrote, may be clearly demonstrated—"the peripheral or radiate archetype, the articulate or longitudinal archetype, the massive or molluscous archetype, and the archetype of the Vertebrate."[47] Since the type was a product of development, "the type itself never exists pure, but only under certain modifications."[48]

Huxley applied von Baer's methodology extensively to his own work in invertebrate morphology and in evaluating other scientists' research.[49] By "observing the 'habits' of living bodies, their mode of development and generation," he hoped to be able to establish the relationships of organisms to one another.[50] In particular, he directed his attention toward Cuvier's radiata, which he described as a "sort of zoological lumber room."[51] As a result of his investigations, he formed several new classes of the radiata. At the same time he demonstrated that certain groups which Cuvier had identified as distinct classes were related. He hoped, he wrote his sister, that his paper "On the Anatomy and the Affinities of the Family of the Medusae" would

> achieve one of the great ends of Zoology and Anatomy, viz. the reduction of two or three apparent widely separate and incongruous groups into modifications of the single type, every step of the reasoning being based upon anatomical facts.[52]

Following von Baer's methodology and using the type concept as an organizing principle, Huxley demonstrated the interelatedness of a widely disparate group of organisms. He discovered that the organs of the medusa were made out of two distinct membranes—the foundation membranes. He believed that these membranes were "one of the essential peculiarities of their structure and that a knowledge of the fact is of great importance in investigating their homologies."[53] Tracing the embryological development of these two membranes in order to identify homologies and also using von Baer's concept of gradation, Huxley extended his investigations beyond the medusa to the whole group which later became known as the Coelenterata. By following the development of two widely differing types, the polyps and medusa, he showed that they were all related. He demonstrated that five widely disparate families of organisms (the Medusae, Physophoridae, Diphydae, Sertularidae, and Hydrae) were not distinct but were "members of one great group organized upon one simple and uniform plan, and even in their most complex and aberrant forms, reducible to the same type."[54]

The mollusks were another group in total disarray. In his paper "On the Morphology of the Cephalous Mollusca," Huxley employed von Baerian principles to see "how far the known laws of development account for these forms, and thence of what archetypal form they may be supposed to be modifications."[55] After describing the general anatomical characters of the group, Huxley determined the cephalous mollusca archetype by studying the embryological development of a variety of its members. Many of the organisms in the group widely differed. Nevertheless, he demonstrated that the "cephalopoda and gasteropoda are morphologically one, are modifications of the same archetypal mollusca form."[56] By carefully comparing many different types of organisms within the group, he described the general characteristics of the mollusca archetype:

> The Archetype of the Cephalous Mollusca then, it may be said, has a bilaterally symmetrical head and body. The latter possesses on its neural surface a peculiar locomotive appendage, the foot; which consists of three portions from before backwards, viz. the propodium, the mesopodium and the metapodium, and bears upon its lateral surface a peculiar expansion, the epipodium.[57]

In determining the molluscan archetype Huxley distinguished it from the other archetypes: "In the cephalous mollusca it is the haemal side of the body which is first developed. In the Articulata and Vertebrata it is the neural side which first makes its appearance."[58] Huxley's research indicated that the types were distinct with no possibility of transition between them. While he said that unmodified forms of the mollusca, vertebrate, and articulate archetypes essentially corresponded, "nevertheless the differences between [*sic*] the three archetypes are so sharp and marked as to allow no real transition between them."[59] The archetype of the cephalous mollusca "in all its modifications is sharply separated from other archetypes whatever apparent resemblances or transition may exist."[60]

Although he believed that the archetypes were distinct, Huxley realized that von Baer's work implied a unity of plan in animal organization, even across the different *types*. "An insect is not a vertebrate animal, nor are its legs free of ribs. A cuttle fish is not a vertebrate animal doubled up."[61] Geoffroy St. Hillaire in his advocacy of transmutation tried to demonstrate that a morphological continuity existed from mollusks to vertebrates. Bending a quadruped backwards, he claimed, resulted in an arrangement of organs similar to that of a cuttlefish. But Huxley sided with Cuvier and von Baer, who argued for the distinctness of types. Nevertheless, he recognized "there was a period in the development of each when insect, cuttlefish and vertebrate were indistinguishable and had a *common plan*."[62] Von Baer believed that the embryo of a vertebrate was from the very beginning a vertebrate animal. However, he admitted that no permanent animal form existed that had not undergone some morphological differentiation during its embryological stages. The further back one traced development, the more similar widely different animals appeared. "Are not all animals essentially similar at the commencement of their development—have they not all a common primary form?" von Baer asked.[63] He suggested that if it was possible to go far enough back in development, one might even attain a state in which the embryos of the invertebrates were indistinguishable from those of the vertebrates.[64] Huxley took the idea even further, claiming that one could trace the "absolute identity of plans" of animals and plants. The plant, he wrote, was "an animal confined in a wooden case."[65]

Huxley's linking of plants to animals by way of development might appear to be an evolutionary explanation for their similarity, but it was not. Organisms might follow a similar plan of development and hence appear identical in the very earliest stages of development, but they were *not* the same. Each type was distinct: the similarity of appearance was not due to common descent. Still, as early as 1853 Huxley recognized the potential interrelatedness of all animals and perhaps plants as well, although he remained committed to a concept of type. Nevertheless, if he were to accept a theory of transmutation, he had somehow to reconcile the two ideas: interrelatedness versus absence of transitional forms. How can organisms be related and still be organized according to distinct types?

In a review article about the cell theory recently put forward by Schleiden and Schwann, Huxley asked, "What is the meaning of the unquestionable fact that the first indication of vitality in the higher organism at any rate, is the assumption of the cellular structure?"[66] Huxley answered his own question by saying that we must look to von Baer and the nature of development for the answer. Von Baer tells us that the "history of development is the history of a gradually increasing differentiation of that which was at first homogeneous."[67] For Huxley, von Baer's ideas continued to be the guiding principles in his own work throughout the 1850s. Thus, in his 1856 lectures on general natural history, Huxley said,

> For animals . . . like all living beings not only *are*, but *become*.
> . . . Before we can affirm that two animals are constructed
> upon a common plan or that two parts are homologous
> (which simply means that they are modifications of corre-

sponding members of a common plan) we must be able to show that these parts or these animals have passed through a corresponding series of developmental stages. It is the absence of this reference to development which is the vice of the ordinary works.[68]

Huxley primarily used von Baerian principles in his own work, that is, he used embryological criteria to determine the relationship of organisms to one another. Nevertheless, his insistence that there were not any and could not be any transitional forms between types appears to be due primarily to the influence of Cuvier, whom he cited as an authority on the issue.[69]

Although von Baer was influenced by *Naturphilosophie*, and in spite of the differences between his approach and Cuvier's, their work had much in common, and what they shared was what was particularly attractive to Huxley. Cuvier, like Huxley, had no use for the "transcendentalists." Cuvier had demolished the theory of the *Echelle des êtres*, which classified organisms in a linear chain from simple to complex by delineating four distinct plans of animal structure. Furthermore, it was not possible to arrange animals in a linear series, even within a single *embranchement*. Von Baer came to the same conclusion. It might be possible to determine a serial arrangement of single organs from the simplest to the most complex, but organisms followed many different paths of development in adapting to different environmental conditions. Both Cuvier and von Baer clearly demonstrated that individual species within a given type might be highly developed with respect to one organ system and quite primitive in all the others. The venerable *scala naturae* simply did not accurately represent the natural order.[70]

As I have mentioned, Huxley did not believe that functional adaptation was what had primarily guided Cuvier in his classifications. Furthermore, such an interpretation suggests that Cuvier recognized that structure may limit the amount of functional adaptation possible. But we do not have to rely on

Huxley's judgment to see that both von Baer and Cuvier saw form and function as intimately linked. Von Baer regarded the material constitution of bodies as the essential determinant of functions. He believed that it was necessary to understand the chemical composition of organisms. He did not advocate absolute physical reductionism, maintaining that chemical processes within an organized body were different from those in the laboratory. Nevertheless, von Baer did not think this meant that a supra material vital force was needed to explain these altered chemical processes. Rather, the vital forces lay in the relation of organic parts with respect to one another. Morphology was the product of the interdependent relationship between form and function. Von Baer, like Cuvier, treated the organism as a functional whole. The position and arrangement of organs were the expression of functional laws of organisms. Thus, these laws were responsible for the forms of various organisms, and the key to understanding the relationships of the various forms could be found in the empirical observation of embryological development. Cuvier also acknowledged the interrelationship of morphology to physiology. Although Cuvier primarily studied the structure of adult organisms, he also recognized the importance of embryology. He studied the fetal membrane of mammals to try to determine their homologies and used embryological arguments to criticize the vertebral theory of the skull.[71] Determining homologies, affinities, and analogies of the various organ systems between different organisms was the cornerstone of classification. These problems were the focus of Huxley's early research.

The Meaning of Analogy and Affinity

Huxley's early work superficially appears to be merely a series of detailed monographs on various invertebrates. But, as indicated in the introduction, he had a much more ambitious research program. Unpublished notes from 1846–1847 document that he was struggling to provide a theoretical foundation for taxonomy.[72] Crucial to this enterprise was the necessity of providing a rigorous definition of the meaning of affinity and analogy. These two terms would describe the fundamental relationship of organisms to one another. William Sharp MacLeay was an important influence in Huxley's early thinking about the matter.

MacLeay was an amateur entomologist and the author of a system of classification known as quinarianism, or the "circular system." Although not well known, he gained some publicity (or notoriety) when his ideas were praised in the anonymously published *Vestiges of Creation* in 1844. It is not clear exactly when Huxley met MacLeay, but it was sometime during the period he was serving on the HMS *Rattlesnake* (1846–1850), which periodically docked in Australia.[73] Huxley always acknowledged MacLeay in his early work.[74] Huxley's first publications dealt with the dissection and classification of marine invertebrates, primarily the Radiata and what became known as coelenterata—although he proposed the term "nematophora." As a result of this work he noted basic similarities between medusae, siphonophores, and hydroids. He claimed that the various forms in these groups were modifications of two primary types—"one series starting from the Anthozoid form of polyp and the other from that of the Hydra—and both running through a parallel and strictly equivalent set of modifications."[75] Although his classificatory scheme appears to be the result of his own research, it was quite similar to the ideas of MacLeay. He wrote to zoologist Edward Forbes, "I was astonished to find how closely some of my own conclusions had approached [MacLeay's], obtained many years ago in a perfectly different way. I believe that there is a great law hidden in the 'circular system' if one could but get at it, perhaps in Quinarianism too."[76]

Essential to MacLeay's system were three ideas: (1) Natural affinities, or relations of closest similarity, lead from one form to the next in a linear fashion. (2) Such series of affinity may run parallel to one another, the parallelism being established by connections, like rungs of a ladder, linking each member of one series across to the corresponding member of the other series. (3) In any natural group, the series of affinities may be represented by a circle. That is, if one followed a series say from one to five, the next link would be from five to one.

MacLeay defined affinity as the close tie linking members of one circular series; analogy was the relationship between corresponding members of a parallel series. This was similar to Huxley's ideas, although he did not represent his series as a circle. Affinities connected group to group in a linear fashion. Correspondences existed between members of one series and the members of a parallel series. During his friendship with MacLeay, Huxley developed his own ideas about the meaning of affinity and analogy. While Huxley believed that searching for analogies between groups was valuable, he also thought that until the existence of analogies had become an established law of zoology, a supposed analogy could not be used as evidence to support the formation of groups.[77] Huxley was searching for a law that explained affinity and analogy. Influenced by von Baer, he believed that the key to the distinction between the two lay in development. Rather than traditional comparative anatomy, one needed to do comparative embryology to find a natural classificatory scheme. Huxley's definition of affinity was actually quite different from MacLeay's. For MacLeay, the analogy between parallel orders was based on their mode of development. The two kinds of resemblance were simply in the pattern of their arrangement. Affinity was the relationship that existed between members of the same series, while analogy was the relationship that existed between members of the parallel series. However, Huxley astutely realized that the two terms in the final analysis were indistinguishable. The same character might be considered an analogy in one case and affinity in another depending on whether they were in the same or different series.

Most naturalists defined affinity as a strong similarity in important characters and analogy as a peculiar similarity between species not related by affinity. But this was too vague. Huxley carefully defined what he meant by these terms because he believed that "the establishment of affinities among animals has been so often a mere exercise in imagination."[78] Furthermore, to demonstrate a real affinity between different classes, it was not enough to show certain analogies or similarities, but that they must be constructed on the same anatomical type, the organs must be shown to be homologous.

Huxley was searching for a law that explained affinity and analogy. Why was this so important? Because, he believed, this was the answer to an even more general question, "What is the end of zoologic science?" which, for Huxley, was to "define the doctrine of animal form." The purpose of classification was to "throw our knowledge of the isolated individual varieties of form into the fewest possible and most general propositions." Huxley could cite several other scientists who supported this view: from Jacob Schleiden's work on botany, "the proper scientific aim of classification is to obtain a complete view of all morphological processes so far as plants proceed from them"; to William Swainson, "The merits of a natural system are in proportion to the number and universality of the facts which it can explain by certain general laws."[79] A clear understanding of affinity and analogy, Huxley believed, was the key to finding these general laws that everyone was talking about. He wanted to develop a general definition that was sufficiently rigorous that it could be applied in a practical way.

Naturalists clearly thought that affinity and analogy referred to two distinct types of resemblances. As far as Huxley could tell, affinity was a close and therefore important resemblance, while analogy was more distant and therefore unimportant. But how did one determine which characters were important and which were unimportant? He thought unimportant ones were essentially isolated—"they involve nothing else and unify nothing else." He gave as an example similarity of color. How-

ever, the dental and ostelogical similarities of the higher vertebrates were important resemblances because more general
propositions could be drawn from them. Important resemblances or affinities were those which were invariably found to
be signs or marks of the presence of other resemblances. All
other resemblances, Huxley claimed, were analogies.

In reviewing the work of earlier naturalists, he noted that
some privileged external characteristics while others used
internal features of anatomy in constructing a classification.
Some thought adult structures were more important than
embryological ones in determining relatedness. But most naturalists agreed that a variety of approaches had to be used.
Huxley concluded that affinity could be indicated by (1) resemblances of external structure; (2) resemblances of internal structure; or (3) resemblances of development. Right away, he saw a
problem. What if animals agreed in structure but differed in
development, or the reverse—agreed in development but differed in the final structure? Which should take precedence?
Again Huxley clarified the terms he was using. "Animals resemble one another in development—*When their various organs
have a similar law of growth*" (emphasis in original). Structural
similarity meant something quite different. If two organs were
said to be the same, this did not mean that they were identical
or even necessarily that they had an outward resemblance. For
example, "The arm of a man and the foreleg of a horse are the
same organs—but they by no means resemble one another in
external appearance." Why? Because, no matter how different
they might look, the two forms could be considered the same
or were said to share an affinity because "*there exists a series of
intermediate forms—the difference between any two of which is
not greater than might be accounted for by a (hypothetical) law of
growth pervading the whole series*" (emphasis in original).[80] In
the case of the forelimb of a horse and the arm of man, there
existed a series of forms such as the forearm of the monkey and
the lemur, the forefoot of the bear, and so on, by which the

forearm could be transformed conceptually into the other. The difference could be the result of laws of consolidation and changes of position. These changes could be considered lawful, Huxley believed, because he could demonstrate such consolidation had occurred in a variety of other cases. Summing up then, organs were considered to be the same or homologous, even when they appeared structurally different, when they shared a resemblance in light of the laws of growth. These laws of growth were at this point hypothetical, Huxley admitted, but they were not arbitrary because they were similar to ones already known—that is, to patterns that had been demonstrated by observation.

In defining affinity as he did, it becomes apparent why Huxley regarded von Baer so highly. Von Baer provided a practical way of determining the relationships between various organisms. By studying their patterns of development, one could empirically demonstrate how forms could be transformed into one another. Not only had von Baer shown that there was a distinct pattern of development for each of Cuvier's *embranchements*, but by observing the larval forms of such anomalous groups as cirripedes, parasitic copepods, and *Comatula*, he had attempted to clarify their relationship.[81] Von Baerian techniques, however, were more than just a methodology for determining relatedness. Huxley was arguing that where the relationship was defined as an affinity or a homology, because one of the following existed—(1) identity of structure, (2) resemblance in external structure, (3) resemblance of internal structure, and (4) resemblance of development—in all these cases what was implied was a similarity in the laws of growth. Furthermore, if two forms were similar, even identical, they were merely analogous if they had come about by different processes of growth. Thus, the free-swimming capsules of hydroids described by Félix Dujardin were analogous to the true medusae. Although they looked similar, medusae had a different developmental pattern than the hydroids.

Huxley cited the science of philology in support of his approach. Words were related by affinity or analogy depending not just on their similarity, but also on whether they could be traced back to the same or different words of origin.

> Philology demonstrates that the words are the same by a reference to the independently ascertained laws of change and substitution for the letters of corresponding words, in the Indo-Germanic tongues: by showing in fact, that though these words are not the same, yet they are modifications by known developmental laws of the same root.[82]

Huxley's distinction between affinity and analogy was quite similar to one that had been made between homology and analogy by a man who was later to become his archenemy— Richard Owen. In 1854, however, Huxley praised Owen and credited him with "the most elaborate and logical development of the doctrine."[83] Owen defined an analogue as "a part or organ in one animal which has the same function as another part or organ in a different animal" and a homologue as "the same organ in different animals under every variety of form and function."[84] Huxley thought Owen's definition of homologue was tied to his belief in a Platonic archetype, which Huxley despised. But Owen, like von Baer, believed that "there exists doubtless a close general resemblance in the mode of development of homologous parts."[85] It was Owen's definition of homology that was adopted by morphologists, including Huxley.

Classification and Evolutionary Theory

From our present-day evolutionary vantage point, Huxley's distinction between affinity (or homology) and analogy seems to shout out for a classification system based on descent from a common ancestor. But as Winsor has pointed out, we have no evidence that Huxley was interested in finding an evolutionary explanation for morphological relationships. There is no indication in the 1840s or early 1850s that Huxley went beyond the embryological history of an individual (ontogeny) to speculating on the historical development of an entire lineage (phylogeny). His citing of philology in support of his method of determining whether organs were homologous or analogous shows he made no distinction between ontogeny and phylogeny—a crucial step in arriving at an evolutionary view of development. It remained for Darwin to employ von Baer's ideas of type in advancing evolution. Nevertheless, in attempting to define analogy and affinity rigorously, Huxley used reasoning virtually identical to Darwin's in attributing these relationships to descent from a common ancestor.

Darwin's theory was built on the ubiquitousness of variation. Nature would have nothing to select if that variation did not exist. But what were the causes of that variation? In his chapter "Laws of Variation" Darwin noted that the pattern of development was highly regulated to control the overall organization. Thus, when slight variations in one part occurred and were accumulated by natural selection, other parts would become modified as well. These laws of development resulted in important structures being modified "independent of utility and therefore of natural selection." For example, blue-eyed cats are usually deaf, but no one would want to argue that deafness was an adaptation. Likewise, Darwin commented that differences between the ray and central florets in some species of Compositous and Umbelliferous plants were often accompanied by the abortion of parts of the flower, but he did not think this was of any service to the plant.[86] Traits that were important to systematists "may be wholly due to unknown laws of correlated growth, and without being as far as we could see of the slightest service to the species."[87]

While these correlations might not be of service to the organism, they certainly contributed to a unity of type, and

Darwin recognized that the morphological type concept was not incompatible with his theory. As he wrote in the *Origin*, "All organic beings have been formed on two great laws—Unity of Type and the Conditions of Existence." Unity of type was the result of unity of descent and was "quite independent of their habits of life."[88] Like Huxley, he also was trying to distinguish his use of the word "type" from the more mystical connotations of the transcendentalists. He wrote Huxley: "The discovery of the type or 'idea' (in your sense for I detest the word as used by Owen, Agassiz, and Co.) of each great class, I cannot doubt, is one of the very highest ends of natural history."[89] In spite of Huxley and Darwin "detesting" Owen's concept of type, in a practical sense their use of the term "archetype" often differed very little from Owen's. The archetype was arrived at by tracing patterns of development to determine which structures were analogous and which were homologous. Owen had clearly articulated what these two terms meant. No major conceptual shift was needed for taxonomists to incorporate evolutionary perspectives into their work. Owen's definition of homology has stood the test of time precisely because it was compatible with Darwin's theory of descent.

After 1859, morphologists, including Huxley, continued to search for homologies and thereby contributed to a fuller articulation of evolutionary theory. Huxley was an outstanding example of a scientist who maintained a belief in the type concept but eventually interpreted it in terms of descent theory. For him, Darwinism provided the perfect means to free the type concept from transcendental overtones. Although Huxley had a type concept quite distinct from that of the *Naturphilosophen*, he continually had to explain that it had no relationship to the Platonic entity, an ideal form that had no physical reality. Descent theory meant that "unity of plan" could no longer be regarded as a manifestation of divine archetypal ideas. Rather, it had a real historical basis, due to the inheritance from a common ancestor. Huxley used the morphological type concept to work out the details of classification. In doing so he hoped to provide evidence for a classification that was as close to a phylogenetic system as possible. Thoroughly grounded in contemporary biological thought, his belief in type was compatible with evolutionary thinking.

Darwin realized that good naturalists, in their search for affinities, were trying to distinguish superficial similarities from more meaningful ones, although they did not realize that the patterns they sought were due to descent from a common ancestor. The types of evidence they found useful for purposes of classification were the same kinds of evidence that Darwin would use to support his theory. That theory told taxonomists *why* organisms could be arranged in certain ways, but the outlines of that arrangement were well in place pre-*Origin*.[90]

Huxley was interested in discovering those structural facts, believing that "natural classification is . . . the same thing as the accurate generalisation of the facts of form, or the establishment of the empirical laws of correlation of structure."[91] For Huxley, a natural classification was possible only because organisms were constructed according to archetypes:

> [I]t is obvious that if animals and plants were not constructed upon common plans, it would be impossible to throw them into groups expressive of their greater or less degree of resemblance, such as those of the natural classification. In fact the doctrine of "common plan" and of "natural classification" are but two ways of expressing the great truth, that the more closely we examine into the inner nature of living beings, the more clearly do we discern that there is a sort of family resemblance amongst them all, closer between some, more distant between others, but still pervading the whole series.[92]

Huxley acknowledged that *The Origin of Species* "introduced a new element into taxonomy. . . . Phylogeny was no less important than Embryogeny in the determination of the systematic place of an animal." But, he added, "we have so little real knowledge of the phylogeny even of small groups, while that of larger groups of animals we are absolutely ignorant."[93] Although phylogenetic speculations were of "great interest and importance," they were "at the present for the most part incapable of being submitted to any objective test."[94] Thus, for Huxley, evolutionary theory gave meaning to taxonomic relationships, but it did not change how one empirically discovered those relationships. As late as 1876, he argued that classification must remain independent of the theoretical aspects of evolution.

> Taxonomy should be a precise and logical arrangement of verifiable facts; and there is no little danger of throwing science into confusion if the taxonomist allows himself to be influenced by merely speculative considerations. The present essay is an attempt to set a good example, and, to draw up a classification of the animal kingdom without reference to phylogeny.[95]

In spite of Huxley's disclaimer, his morphological work in the late 1860s and 1970s *was* concerned with questions of phylogeny as well as taxonomy. However, he believed the best way to further the cause of evolution was to emphasize the empirical basis of taxonomic relationships. A classification based on "well established facts may have some chance of permanence, in principle, if not in detail, while the successive phylogenetic schemes come and go."[96] Huxley seems to have forgotten that numerous typological schemes had also come and gone, some of which were also quite speculative. And certainly many changes in classification have been made on the basis of phylogenetic considerations. He was naive in his belief that a clear distinction always could be made between "facts" and the interpretation of them. Phylogenies are difficult to determine precisely because of the problem of interpretation. Comparing similarities of structure has been the standard way to determine taxonomic relationships. However, the system of classification known as cladistics has come to dominate the science of systematics by developing a set of procedures to classify organisms based on phylogenetic considerations only. Nevertheless, modern cladists argue vociferously over whether a particular trait is derived or primitive. However, often the difficulty in interpreting the data is because the data is incomplete. Always the empiricist, Huxley envisioned science as a truly Baconian project: Collect enough facts, and the great underlying truths will eventually emerge. Such a view of classification was compatible with evolutionary thinking but at the same time independent of it. It also explains why Huxley's own research program could and did provide support for Darwin's theory without making use of natural selection theory.

Development and the Fossil Record

In the previous section on classification, I argued that taxonomy based on morphological criteria and the type concept may exist independently of evolutionary thinking and still be compatible with evolution. This section examines the relationship of development to the fossil record and the role the type concept played in these discussions. Developmental arguments were used both to support and to discredit transmutation, and Huxley once again was in the middle of these debates. The concept of type underlay von Baer's famous laws of development: (1) that the more general characters of a large group of animals appear earlier in their embryos than the more special characters; (2) from the most general forms the less general are developed, until finally the most special arise; (3) every embryo of a given animal form, instead of passing through other forms, rather becomes separated from them; and (4) fundamentally, therefore, the embryo of a higher animal form is never identical to any other form, but only to its embryo.[97] In the third and fourth laws, von Baer was specifically trying to distinguish his views from the theory of recapitulation, which claimed that the ancestral adult stages are repeated in the embryonic or juvenile stages of descendents. The embryo of a higher animal did not pass through the adult forms of lower animals, but only through the embryonic forms of lower animals within its own archetype. Darwin used the type concept in applying the laws of embryological development to the fossil record. However, Darwin was not the first to draw attention to certain parallels between the sequence of the fossil record and the stages of embryological development. William Carpenter and Richard Owen also commented on these similarities, but whereas Darwin used this information to bolster his theory of evolution, Carpenter used the "laws of development" to argue explicitly against transmutation.

Borrowing von Baer's idea that the undifferentiated germ became specialized as development proceeded, Carpenter applied it to the fossil record. In his *Principles of Physiology*, he suggested that the fossil record showed a progression from a generalized archetype to more specialized forms. He acknowledged that the history of the earth showed a "series of organic lines to have successively appeared and disappeared on the face of the globe."[98] But as to the suggestion that an actual transmutation of the lower forms into the higher took place, he found very little evidence. He only grudgingly accepted the idea that the Radiata, Mollusca, and Articulata had *perhaps* existed before any vertebrated animals left traces of their existence.[99] Even if the doctrine of progressive development were true, he wrote,

> it would afford no ground whatever for the doctrine of *trans-mutation*, which is not only opposed to all our experience, but which fails to account for the intimate *nexus* that so commonly unites together, not merely the higher and the lower forms of each series, but the members of different series with each other.

A more satisfactory account of the Succession of Organic Life on the surface of the globe, may probably be found in the general plan which has been shown to prevail in the development of the existing forms of organic structure, namely the passage from the more general to the more special. . . . We find a certain class of cases in which extinct animals, especially the earliest forms of any class that may be newly making its appearance, present indications of a closer conformity to "archetypal generality" than is shown in the existing animals to which they bear the closest approximation; and hence their conformity to the latter is closer in the embryo-condition of these than in their fully developed and more specialized state.[100]

Thus, like von Baer, Carpenter used the pattern of development to argue against transmutation.

Carpenter, Owen, and Agassiz all saw parallels between fetal development and the history of life recorded in the fossil record.[101] Owen applied von Baer's ideas to the interpretation of fossil record and came to the same conclusions as Carpenter, precipitating a minor priority dispute.[102] Louis Agassiz, as early as 1844, had taken a similar position: "[T]he successive creations [of life on the earth] have passed through phases of development analogous to those through which the embryo passes during its growth."[103] But Agassiz and Carpenter had a somewhat different interpretation than von Baer. They argued that the fossilized *adult* forms of extinct species were recapitulated in present-day embryos. Von Baer was clear that the resemblances would be to ancestral embryonic forms, not adult ones. In the 1840s and 1850s leading biologists applied von Baer's ideas to the fossil record, but without arriving at an evolutionary interpretation. The divergence observed both in fetal development and in the fossil sequence seemed to them to be part of a wider developmental plan.

In *The Non-Darwinian Revolution*, Peter Bowler draws a sharp distinction between a developmental and an evolutionary view of the history of life. In this analysis he labels Huxley a "pseudo-Darwinian" who belongs to the tradition of pre-Darwinian developmental biologists.[104] However, the two ideas, development and evolution, are clearly not incompatible. One needs only to look at Darwin's works to see how von Baer's ideas could be applied within an evolutionary framework. If an emphasis on development makes someone a pseudo-Darwinian, then Darwin is a pseudo-Darwinian as well. Huxley was determined to find a way to make a belief in type compatible with a theory of transmutation.

Linking embryology to paleontology had great appeal for Darwin: "embryology is to me by far the strongest single class of facts in favour of change of forms."[105] In his treatment of phylogeny he was influenced by von Baer and Carpenter as well as Owen and Agassiz. Darwin's ancient progenitors were the archetypes of extant animal species. He believed that in these ancient animals the adult form and the embryo were similar. The archetype was in some degree embryonic and therefore capable of undergoing further development, he wrote Huxley.[106] Thus, even in modern, highly developed species the embryo still had the general unmodified form of the archetype. For von Baer, these resemblances were the necessary consequence of development from a common starting point by a single process of increasing specialization. For Darwin, the pattern in the fossil record was the result of descent from a common ancestor with divergence and increasing specialization occurring over time. The variations from the general archetype were inherited, but they usually did not make their appearance until late in development, the embryonic stages remaining unchanged. "The embryonic state of each species and group of species partially shows us the structure of their less modified ancient progenitors."[107] For Darwin, the archetype was a real organism that had existed in the distant past. Far from thinking that the type concept was contrary to evolutionary thinking, Darwin found it useful in dealing with embryology and the fossil record.[108]

In the nineteenth century many who had a developmental view argued explicitly against Darwin. Nevertheless, it is important to recognize that in arguing against Darwin's theory, developmentalists were not just misunderstanding the theory or tied to backward ideas. Rather, they raised issues that evolutionists would have to address, issues that remain problematic today and reflect an ongoing tension between two different research programs in trying to understand the structure of organisms. Which is primary: form or function? Von Baer was a case in point. In arguing against "usefulness" as a criterion for evolutionary change, von Baer pointed out that there would have been many times in evolutionary history that having four legs and two wings would have been advantageous, yet such a form never evolved.[109] He believed that internal factors limited the types of variability produced that natural selection could act on. Because of these constraints, natural selection could not account for the origin of von Baer's distinct types from a single primeval germ. Interestingly enough, von Baer accused the Darwinians of being guilty of the same type of thinking as the *Naturphilosophen*: By arguing for the unity of descent from a single ancestral form, the Darwinians were no different from the transcendentalists, who believed in an absolute ideal form, since no direct empirical support existed for either claim.[110]

The morphological tradition has been accused of not having a theory or that the theory was vacuous because it did not explain unity of type. But as Ron Amundson correctly points out, the transcendental morphologists were the ones to call attention to the unity of type in the first place.[111] In their attempt to unify the organic world, they developed the concept of homology, which has been and remains crucial to evolutionary classification. In addition, many of the transcendental morphologists used typological theories as evidence to argue against the creationism of the natural theologians. Furthermore, while Darwin credited "the illustrious Cuvier" with the idea that the "conditions of existence" were responsible for producing adap-

tation, Huxley has convincingly shown that Cuvier's classification was based on morphological consistency, rather than on
functional analysis, in spite of always being cited as the arche-
typical example of a functionalist. Likewise, Richard Owen was
often referred to as the British Cuvier. However, in his recent
biography of Owen, Nicolaas Rupke argues persuasively that this
label was inappropriate because by the end of the 1840s Owen
believed that form was the key to solving the problem of organic
diversity.[112] Not only is there a theoretical debate over whether
form determines function or visa versa, but there is a practical
one as well. Perhaps studying how form comes to be generated,
that is, by studying development, is a better research strategy to
understanding function and adaptation. Huxley clearly thought
this was the case, and thus his role in the debate between the
developmentalists and the Darwinians was unique because he
managed to be in both camps simultaneously. These debates also
show that in order to properly assess Darwinism's position in
nineteenth-century biology, it must be regarded as comprising
more than the theory of natural selection.

Certainly many scholars have acknowledged the many
aspects of Darwinism.[113] Nevertheless, much of the criticism of
Huxley seems to be directed at his lack of enthusiasm for natural selection. Bowler argues in *The Non-Darwinian Revolution* that in the nineteenth century a developmental view pre-
vailed. Since what he considers the most important aspects of
Darwin's theory (natural selection and a nonteleological view
of adaptation) were not well accepted, why should we call it a
revolution?[114] While I disagree with this whiggish analysis, he is
certainly correct that natural selection was not well accepted.
But it is precisely because Darwin's theory does encompass
much more than just natural selection that its impact was revo-
lutionary, both in its implications as a "worldview" as well as in
its providing a research program for virtually all aspects of
biology: embryology, anatomy, taxonomy, paleontology, bio-
geography, and heredity. Huxley recognized this:

I by no means suppose that the transmutation hypothesis is proven or anything like it. But I view it as a powerful instrument of research. Follow it out and it will lead somewhere; while the other notion [special creation] is like all modifications of "final causation": a barren virgin.[115]

The line between the developmentalists and evolutionists is not a sharp one. Darwin certainly recognized the importance of development for his theory. However, Huxley's early research of the 1840s and 1850s in developmental morphology led him to adopt a typological species concept, which resulted in his arguing against transmutation. Along with Cuvier and von Baer, he had no use for the evolutionary theories of Lamarck, Geoffrey, or Robert Chambers. Indeed, when Darwin's theory came along, Huxley raised many of the same objections as von Baer. Like von Baer, he doubted whether natural selection was a powerful enough mechanism to cause species change, and in addition he also disagreed with Darwin's claim that change was slow and gradual. Yet he accepted Darwin's theory, and von Baer did not. But Huxley differed from von Baer and most of the developmentalists in one very important respect. And it was this difference which allowed Huxley to enthusiastically embrace Darwin's theory in spite of his problems with it. To borrow Lenoir's terminology, von Baer was a teleomechanist; that is, teleology was fundamental to his theory of development and was ultimately responsible for his rejection of Darwinism.[116] But Huxley had no attachment to teleology. In fact, it was precisely because Darwin's explanation was free from any notions of final causes that he found it so attractive. In addition, Darwin's theory explained an enormous number of "facts," making it irresistible to the empirically minded Huxley.

Huxley's Dilemma

Working in the tradition of Cuvier and von Baer, Huxley found powerful support for the type concept. As Agassiz wrote of the first two men's work,

> Here then was a double confirmation of the distinct circum-scription of types, as based upon structure, announced almost simultaneously by two independent investigators, ignorant of each others' work, arriving at the same result by different methods. The one . . . in the embryonic germs of various animals worked out the four great types of organic life from the beginning; while his coworker reached the same end through a study of their perfected structure in adult forms.[117]

While von Baer was certainly familiar with Cuvier's work, in spite of Agassiz's claim, this does not invalidate his main point: Two very different approaches toward the study of natural history had given similar results. How Huxley reconciled the absence of transitional forms with a theory of transmutation is the subject of the next chapter.

Notes

1. Thomas Huxley, November 23, 1859, *Life and Letters of T. H. Huxley* (*LLTHH*), 2 vols., Leonard Huxley, ed. (New York: D. Appleton & Co., 1901), 1: 189.

2. Thomas Huxley, 1860, "The Origin of Species," in *Collected Essays of Thomas Huxley*, vol. 2, *Darwiniana* (London: Macmillan & Co., 1893), p. 77.

3. Thomas Huxley, 1863, "On Our Knowledge of the Causes of the Phenomena of Organic Nature," in *Darwiniana*, pp. 475–76.

4. Thomas Huxley to George Romanes, May 9, 1882, *LLTHH* 2: 42.

5. Timothy Lenoir, *The Strategy of Life* (Dordrecht, The Netherlands: Reidel Press, 1982), p. 3.

6. Ken Caneva, "Teleology with Regret," *Annals of Science* 47 (1990): 291–300.

7. Lenoir, *Strategy of Life*, p. 103.

8. Ernst Mayr, *The Growth of Biological Thought* (Cambridge, Mass.: Harvard University Press, 1882), p. 487.

9. Ibid.

10. In more recent writings Mayr has acknowledged that there are different usages of the word "type" and "does not question the propriety of recognizing as purely descriptive the major archetypes or organisms reflected in the taxonomically recognized 27 or so phyla of animals" (in *Towards a New Philosophy of Biology* [Cambridge, Mass.: Harvard University Press, 1988], p. 406). Nevertheless, in his earlier writings, the distinction between this usage and his often vehement criticism of "essentialism" is not at all apparent.

11. William Whewell, *Philosophy of the Inductive Sciences* (London: Parker, 1840), quoted in Paul Farber, "The Type-Concept in Zoology," *Journal of Historical Biology* 9 (1976): 94, n 2.

12. Farber, "Type-Concept in Zoology," pp. 96–97.

13. Ibid., p. 117. In addition to Farber's article, for more information on the position of the type concept in nineteenth-century biological thought see William Coleman, "Morphology between Type Concept and Descent Theory," *Journal of the History of Medicine* 31 (1976): 149–75; Philip Rehbock, *The Philosophical Naturalists* (Madison: University of Wisconsin Press, 1983); E. S. Russell, *Form and Function: A Contribution to the History of Animal Morphology* (1916; Chicago: University of Chicago Press, 1982); and Ron Amundson, "Typology Reconsidered: Two Doctrines on the History of Evolutionary Biology," *Biology and Philosophy* 13 (1998): 153–77.

14. See, however, Nicolaas A. Rupke, "Richard Owen's Vertebrate Archetype," *Isis* 84, no. 2 (1993): 231–25, for a different view of Owen's archetype. He claims that Owen's archtype was definitely not Platonic and even suggests it represented a more Aristotelian view of form.

15. Thomas Huxley, "On the Theory of the Vertebrate Skull," *Proceedings of the Royal Society* 9 (1857–1859): 381–457, in *Scientific Memoirs of Thomas Henry Huxley* (*SMTHH*), 4 vols., Michael Foster and E. Ray Lancaster, eds. (London: Macmillan & Co., 1898–1902), 1: 584–85.

16. Ibid., p. 571.

17. Thomas Huxley, "Natural History as Knowledge, Discipline, and Power," *Royal Institute Proceedings* 2 (1856): 187–95, *SMTHH* 1: 306.

18. Thomas Huxley, "On the Morphology of the Cephalous Mollusca," *Philosophical Transactions of the Royal Society* 143 (1853): pt. 1, pp. 29–66, *SMTHH* 1: 192.

19. Ibid., p. 177.

20. Ibid.

21. Thomas Huxley, "Researches into the Structure of the Ascidians," *British Association Report* (1852): pt. 2, pp. 76–77, *SMTHH* 1: 194. It should be noted that often, as in this case, this paper was not actually written by Huxley. Rather, it was a summary by someone who had attended his lecture. Thus, this represents a quote by another person drawing attention to a point that presumably Huxley emphasized.

22. For more of a discussion on the idealist conception of archetype see Nicolaas Rupke, *Richard Owen, Victorian Naturalist* (New Haven: Yale University Press, 1994), pp. 188–96.

23. Russell, *Form and Function*, p. 3.

24. William Coleman, *George Cuvier, Zoologist* (Cambridge, Mass.: Harvard University Press, 1964), p. 2.

25. Huxley, "Natural History as Knowledge," *SMTHH* 1: 307.

26. Toby Appel, *The Cuvier-Geoffroy Debate* (New York: Oxford University Press, 1987), p. 41.

27. Coleman, *George Cuvier*, p. 100.

28. George Cuvier, *Le Règne Animal* 1:57, quoted in Appel, *Cuvier-Geoffrey Debate*, p. 45.

29. Coleman, *George Cuvier*, p. 101.

30. Ibid., p. 102.

31. Appel, *Cuvier-Geoffrey Debate*, p. 45.

32. Farber, "Type-Concept in Zoology," p. 102.

33. George Cuvier, "Sur un Nouveau Rapproachment," quoted in Appel, *Cuvier-Geoffrey Debate*, p. 45.

34. Huxley, "Natural History as Knowledge," *SMTHH* 1: 308.

35. Ibid., Huxley quoting Cuvier, pp. 308–309.

36. Ibid., p. 309.

37. Ibid.

38. Ibid., p. 310.

39. Coleman, *George Cuvier*, p. 103.

40. Thomas Huxley, 1857, Darwin Papers 205.5, quoted in M. di Gregorio, *T. H. Huxley's Place in Natural Science* (New Haven: Yale University Press, 1984), p. 112.

41. Charles Darwin, *The Origin of Species* (New York: Avenel, 1976), p. 205.

42. Thomas Huxley quoting Cuvier, 1853, "On the Morphology of the Cephalous Mollusca," in *SMTHH* 1: 191.

43. Ibid.

44. Karl Ernst von Baer, 1828, "Philosophical Fragments *Über Entwickelungsgeschichte The Fifth Scholium*," in *Scientific Memoirs, Natural History*, A. Henfrey and Thomas Huxley, eds. (London: Taylor and Francis, 1853), pp. 186–238.

45. Ibid., p. 196.

46. Ibid., pp. 195–96.

47. Ibid., p. 197.

48. Ibid., p. 219.

49. Although I will be citing a few specific instances, a quick perusal through volume 1 of Huxley's *Scientific Memoirs* documents Huxley's use of von Baer in virtually all his articles on invertebrates.

50. Thomas Huxley, December 10, 1846, *T. H. Huxley's Diary of the Voyage of HMS* Rattlesnake, Julian Huxley, ed. (London: Chatto and Windus, 1935), pp. 8–9.

51. Thomas Huxley to Sir John Richardson, October 31, 1850, *LLTHH* 1: 63.

52. Thomas Huxley to his sister Lizzie, August 1, 1847, *LLTHH* 1: 36.

53. Ibid., p. 11.

54. Thomas Huxley, "On the Anatomy and the Affinities of the Family of the Medusae," *Philosophical Transactions of the Royal Society* (1849), pt. ii: 413, *SMTHH* 1: 28.

55. Thomas Huxley, 1853, "On the Morphology of the Cephalous Mollusca," *SMTHH* 1: 153.

56. Ibid., p. 190.

57. Ibid., p. 177.

58. Ibid. "Haemal" refers to the side the heart is on, while the neural side is the side where the nervous ganglion appears.

59. Ibid.

60. Ibid., p. 191.

61. Thomas Huxley, "Abstract on The Common Plan of Animal Forms," *Royal Institute Proceedings* 1 (1851–1854): 444–46, *SMTHH* 1: 283.

62. Ibid.

63. Von Baer, "Philosophical Fragments," p. 213.

64. Ibid., p. 212.

65. Thomas Huxley, 1853, "On the Identity of Structure of Plants and Animals," *Proceedings of the Royal Institute* 1 (1851–1854): 298–302, *SMTHH* 1: 218.

66. Thomas Huxley, "The Cell Theory," *British and Foreign Medical Chiralogical Review* 41 (1853): 285–314, *SMTHH* 1: 264.

67. Ibid., Huxley quoting von Baer.

68. Thomas Huxley, 1856, "Lectures on General Natural History," quoted in di Gregorio, *T. H. Huxley's Place*, p. 31.

69. Huxley, "On the Morphology of the Cephalous Mollusca."

70. For a more complete discussion on the similarities of Cuvier and von Baer in relation to the *scala naturae*, see Russell, *Form and Function*, pp. 128–32.

71. Ibid., pp. 130–31.

72. Thomas Huxley, "Some Considerations upon the Meaning of the Terms Analogy and Affinity," 1846–1847, Huxley Manuscripts (HM) 37:1.

73. Huxley, *T. H. Huxley's Diary of the Voyage of HMS* Rattlesnake.

74. See *SMTHH*, "On the Anatomy and the Affinities of the Family of the Medusae," *Philosophical Transactions of the Royal Society* 1 (1849): 28–29; "Notes on Medusa and Polypes," 1 (1849): 34; "Zoological Notes and Observations Made on Board HMS *Rattlesnake* During the Years 1846–1850," *Annals and Magazine of Natural History* 1 (1851): 82; "An Account of Researches into the Anatomy of the Acalepahe," *British Association Report* 1 (July 1851): 100.

75. Thomas Huxley, October 1849, "Notes on Medusae and Polypes." For a complete discussion of Huxley's classification system and its relationship to MacLeay's, see Mary Winsor's *Starfish, Jellyfish, and the Order of Life* (New Haven: Yale University Press, 1976), pp. 61–97.

76. Huxley, "Notes on Meduse and Polypes," *SMTHH* 1: 34.

77. Winsor, *Starfish, Jellyfish*, p. 91.

78. Huxley, "On the Anatomy and the Affinities of the Family of the Medusae," *SMTHH* 1: 9, n. 24.

79. Thomas Huxley, "Some Considerations upon the Meaning of the Terms Analogy and Affinity."

80. Ibid.

81. He actually mixed up the order, but this does not invalidate the notion that this was a valuable technique for classification.

82. Thomas Huxley, 1854, "Abstract on the Common Plan of Animal Forms," *SMTHH* 1: 283.

83. Ibid., p. 282.

84. Richard Owen, 1846, *Report on the Archetype and Homologies of the Vertebrate Skeleton*, British Association for the Advancement of Science *Report* (n.p.), p. 175. Owen originally defined these terms in 1843 in his *Lectures on Invertebrate Animals*.

85. Richard Owen, 1849, *On the Nature of Limbs* (London: n.p.), p. 6.

86. Darwin, *Origin of Species,* p. 184.

87. Ibid., p. 185.

88. Ibid., p. 233.

89. Charles Darwin, April 23, 1854, *More Letters of Charles Darwin* (*MLCD*), 2 vols. (London: Macmillan & Co., 1902), 1: 73.

90. Mary Winsor, "The Impact of Darwinism upon the Linnaean Enterprise, with Special Reference to the Work of T. H. Huxley," in *Contemporary Perspectives on Linnaeus*, John Weinstock, ed. (Lanham, Md.: University Press of America, 1985), p. 60.

91. Thomas Huxley, "Owen's Place in the History of Anatomic Science," in *The Life of Richard Owen*, 2 vols., Richard Owen, ed. (New York: D. Appleton & Co., 1894), 2: 284.

92. Thomas Huxley, 1865, "Explanatory Preface to the Catalogue of the Paleontological Collection in the Museum of Practical Geology," *SMTHH* 3: 134.

93. Thomas Huxley, "On the Classification of the Animal Kingdom," *Journal of Linnaean Society* 12 (1876), *SMTHH* 4: 36.

94. Ibid.

95. Ibid., p. 37.

96. Ibid.

97. Von Baer, "Philosophical Fragments," pp. 210–14.

98. William Carpenter, *Principles of Physiology* (London: John Churchill, 1853), p. 133.

99. Ibid., p. 134. In other writings, however, Carpenter seems to have been quite an advocate of progressive development.

100. Ibid.

101. Agassiz, however, was developing a theory of recapitulation which was thus different from what von Baer was claiming.

102. For an excellent discussion of the influence of von Baer on the various discussions surrounding the fossil record, see Dov Ospovot, "The Influence of Karl Ernst von Baer's Embryology, 1828–1859: A Reappraisal in Light of Richard Owen and William B. Carpenter's Paleontological Application of von Baer's Law," *Journal of the History of Biology* 9 (1976): 1–28.

103. Louis Agassiz, 1844, "Poisson Fossiles de Vieux Gres Rouge," quoted in ibid., pp. 14–15.

104. Peter Bowler, *The Non-Darwinian Revolution* (Baltimore: Johns Hopkins University Press, 1988), pp. 72–77.

105. Charles Darwin to Asa Gray, September 10, 1860, *Life and Letters of Charles Darwin* (*LLCD*), 2 vols., Francis Darwin, ed. (London: John Murray, 1887), 2: 131.

106. Charles Darwin, April 23, 1854, *MLCD*, 1: 73.

107. Darwin, *Origin of Species*, p. 427.

108. For a provocative discussion of the relationship of embryology to the genesis of Darwin's theory of descent see Robert Richards, *The Meaning of Evolution* (Chicago: University of Chicago Press, 1992).

109. Lenoir, *Strategy of Life*, p. 254.

110. Ibid.

111. Amundson, "Typology Reconsidered."

112. Rupke, *Richard Owen*, pp. 163–72.

113. See John Greene, "Darwinism as World View," *Science Idealogy, and World View* (Berkeley: University of California Press, 1981), pp. 128–57; Frank M. Turner, *Between Science and Religion, the Reaction to Scientific Naturalism in Late Victorian England* (New Haven: Yale University Press, 1974); Ernst Mayr, "Darwin's Five Theories of Evolution," in *The Darwinian Heritage*, David Kohn, ed. (Princeton, N.J.: Princeton University Press, 1985), pp. 755–72.

114. Bowler, *Non-Darwinian Revolution*, chapter 1.

115. Huxley to Charles Lyell, June 25, 1859, *LLTHH* 1: 187.

116. For a good discussion of von Baer's criticism of Darwin's theory see Lenoir, *Strategy of Life*, chapter 6.

117. Louis Agassiz, "Evolution and Permanence of Type," *Atlantic Monthly* 33 (1874): 92–101, in *Darwin and His Critics*, David Hull, ed. (Chicago: University of Chicago Press, 1972), p. 432.

Chapter Three
Lyell, Saltation, and the Species Question

I have claimed that the type-concept was not incompatible with evolution. Nevertheless, many transcendental morphologists who had a typological species concept used that as "evidence" against any theory of transmutation. Cuvier's fossil work indicated that large gaps in the fossil record existed with little evidence of transitional organisms. However, Darwin's evolutionism made a lot of sense to Huxley, causing him to exclaim, "How extremely stupid not to have thought of that!"[1] But Darwin's theory presented a dilemma. How could one reconcile evolution with the concept of distinct types? If transmutation occurred, where were the transitional forms? Huxley's solution to the problem was saltation. Paradoxically, he seems to have come to this view partly as a result of his long discussions with the great promoter of gradualism—Charles Lyell. I will argue that Huxley's early belief in saltation reflected his desire to reconcile his belief in the uniformitarianism of Lyell's geology with the concept of morphological types in biology.[2]

Uniformitarianism

Lyell's influence on Darwin is well known. Most historians have focused on Darwin's adoption of Lyell's gradualist principles and his application of them to the realm of biology. Darwin, of course, was not the only one reading Lyell. Five editions of the *Principles of Geology* appeared in the 1830s; four more were published before 1859. Lyell's influence on his fellow scientists was profound. As Huxley wrote in 1887,

> I cannot but believe that Lyell for others as for myself was the chief agent in smoothing the road for Darwin. For consistent Uniformitarianism postulates evolution as much in the organic as in the inorganic world. The origin of a new species by other than ordinary agencies would be a vastly greater "catastrophe" than any of those which Lyell successfully eliminated from sober geological speculation.[3]

Uniformitarianism had several components to it, but one was of particular importance to Huxley.[4] Uniformitarianism was actualistic. Huxley, like Lyell, wanted a *naturalistic* explanation of earth history. The proponents of catastrophism argued that current geological processes could not explain major geological events such as the destruction or building of mountain ranges. Rather, the earth's surface was shaped by major catastrophic events interspersed with periods of relative stability. Lyell wanted a scientific geology and believed that the catastrophist methodology would prevent geology from rising to the level of an exact science. In the early nineteenth century, catastrophism in Britain was associated with scriptural geology. In other words, findings in geology were used as evidence for the biblical flood. While catastrophism did not require supernatural intervention, this was its legacy. Uniformitarianism was explicit in its denial of supernatural forces to explain earth history, and Huxley heartily approved. For Lyell, a naturalistic explanation meant that the history of life reflected *actualistic* processes. Causes now oper-ating to alter geological formations are the same as those oper-ating in the past. There were two aspects to actualism: causes differ neither in *kind* nor in *degree*. Present-day processes such as erosion, earthquakes, and volcanoes are of the same sort and intensity as those acting in the past. Huxley believed there was no sharp boundary between degree and kind, since it was a matter of definition when a past cause became so different in its energy from its present analog that it deserved to be called dif-ferent in kind. If processes occurring today were basically the same as those occurring in the past, actualism provided a methodology to investigate past events. Huxley had little patience for armchair speculation of what might have happened. Uniformitarianism was not just a theory about earth history; it provided a research program for understanding the past.

Another aspect to Lyell's uniformitarianism was the idea of slow, gradual change. Gradualism is often conflated with actu-alism; however, the two concepts are distinct. The entire his-tory of the earth may be the result of actualistic processes but still be described as episodic or even catastrophic. Periods of rapid, cataclysmic change interspersed with periods of stasis can be actualistic but do not appear to be gradualistic. In the same way, the explanation for species change over time may be actu-alistic, but not gradualistic. Unlike Darwin, Huxley did not apply the gradualism of Lyell's geology to the biological realm. Rather, he was influenced by Lyell's biological views. It is Lyell's views on speciation that I turn to now.

Lyell and the Problem of Species

Historians of science have paid relatively little attention to Lyell's views on speciation,[5] yet Lyell wrote several journals devoted to the species question, and his correspondence with Huxley indicates that they had discussed the question of speciation at great length.[6] Lyell's *Principles of Geology* was not just about the inorganic processes that shaped the earth. Volume 2, known as the biological volume, dealt with changes in the organic world, including those affected by the organisms themselves as well. But historians have shown little interest in Lyell's views on species because he did not believe in transmutation and was only later partly converted by Darwin. Furthermore, his discussion of species origination in the *Principles* was quite vague, referring to the Creator as having brought species into the world by some unknown mechanism. However, it is precisely because Lyell was an antitransmutationist that his thinking about speciation gives us insight into the discussions provoked by Darwin's bold new theory.

As William Coleman pointed out, Lyell recognized that the series of geological strata and the sequence of forms in the fossil record were different aspects of the same problem. The basic question, Lyell declared, was "whether species have a real and permanent existence in nature; or whether they are capable, as some naturalists pretend, of being indefinitely modified in the course of a long series of generations?"[7] To answer this question he presented a detailed analysis of Lamarck's theory of transmutation.

Lamarck defined species as a "collection of similar individuals produced by other individuals like themselves."[8] This definition was incomplete, Lyell claimed, because most naturalists adopted Linnaeus's typological definition: individuals from a common stock shared certain distinguishing characteristics that would never vary and had remained constant since the creation of each species. By the end of his life Linnaeus believed that many present-day species were not part of God's original creation. Rather, later species had arisen by the hybridization of preexisting ones. But this is not the same as claiming that species themselves actually change through time. Influenced by Linnaeus, Lyell thought there was some essence of a species that was unchanging and permanent, even if the individuals that make up a species come and go. However, Lamarck denied that species existed as real entities, claiming instead that only individuals existed. One only had to observe the trouble naturalists had in deciding what was a true species and what was a variety. Summarizing Lamarck's views, Lyell wrote,

> The greater the abundance of natural objects assembled together, the more do we discover proofs that everything passes by insensible shades into something else: that even the more remarkable differences are evanescent, and that nature has for the most part, left us nothing at our disposal for establishing distinctions, save trifling and, in some respects, puerile particularities.[9]

This sounds remarkably similar to what Darwin claimed many years later. However, Lyell denied transmutation of species occurred and attacked Lamarck's explanation of species change.

Lamarck, like Darwin, believed that organisms adapted to their environment. Changes in climate and habitat resulted in a change in the organism's structure. Eventually the alteration became so complete that a new species was formed. How did the change occur? "Every considerable alteration in the local circumstances in which each race of animals exists, causes a change in their wants, and these new wants excite them to new actions and habits."[10] As a result of more frequent use, some parts became more developed, while other parts that were unused atrophied and perhaps disappeared altogether. New parts with new functions replaced unused structures. But Lyell interrupted his presentation of Lamarck's argument and observed "that no positive fact is cited to exemplify the substitution of some *entirely new* sense, faculty, or organ, in the room of some other suppressed as useless."[11] Lyell did not deny that certain faculties may become stronger after many generations of active exertion, or may be enfeebled by generations of disuse. But such processes could not generate new structures. Lyell pointed out to the reader this important chasm in the chain of evidence, because he might otherwise imagine that I had merely omitted the illustrations for the sake of brevity, but the plain truth is that there were no examples to be found; and when Lamarck talks "of the efforts of internal sentiment," "the influence of subtle fluids," and "facts of organization," as causes whereby animals and plants may acquire *new* organs, he substitutes names for things, and with a disregard to the strict rules of induction, resorts to fictions, as ideal as the "plastic virtue," and other phantoms, of the geologists of the middle ages.[12]

Lyell's critique of Lamarck appealed to Huxley for several reasons. First, where were the facts? Huxley wanted concrete examples of what Lamarck was saying. Yet there was no evidence for the *de novo* appearance of new structures. Lamarck had claimed that "the efforts of some short-necked bird to catch fish without wetting himself have, with time and perseverance, given rise to all our herons and long-necked waders."[13] Huxley thought we should be able to observe such processes in action. He granted that organs may be modified by exercise, but it was not apparent that such modifications were heritable. Furthermore,

> It does not seem to have occurred to [Lamarck] to inquire whether there is any reason to believe that there are any limits to the amount of modification producible, or to ask how long an animal is likely to endeavour to gratify an impossible desire. The bird in our example would surely have renounced fish dinners long before it had produced the least effect on leg or neck.[14]

For Huxley, Lamarck's example was implausible for two reasons. As discussed in the previous chapter, a variety of different kinds of evidence indicated that there was a limit to the amount of variability within a specific type. In addition, Huxley's humorous remark about the bird giving up fish dinners made a serious point. An animal was unlikely to maintain an unrewarded behavior pattern for the amount of time necessary to produce the structural change that particular behavior was trying to elicit, even if one accepted that it was possible for acquired characteristics to be inherited. This latter objection related to another aspect of Lyell's critique that impressed Huxley.

Lyell's second objection to Lamarck's mechanism of species change was based on Cuvier's inaccurate portrayal of Lamarck's theory. Cuvier claimed that Lamarck espoused the idea that organisms could "will" a change in their structure, although no such mechanism can be found in Lamarck. For Lyell, the concept of an organism "willing" a longer neck or "contracting a habit" was logically absurd. Rather than positing that certain organs or structures allow the organism to develop certain habits, Lamarck turned the argument on its head. The habits or manner of living eventually determined the form of the animal's body. Lyell described what became Lamarck's most notorious example:

> The camelopard was not gifted with a long flexible neck because it was destined to live in the interior of Africa, where the soil was arid and devoid of herbage; but, being reduced by the nature of that country to support itself on the foliage of lofty trees, it contracted a habit of stretching itself up to reach the high boughs, until its forelegs became longer than the hinder, and its neck so elongated, that it could raise its head to the height of twenty feet above the ground.[15]

The idea of an organism "willing" a change was vague and metaphysical. Although Lamarck had another explanation for transmutation in plants, Huxley ignored it. Rather, he claimed that "effort excited by change of conditions—was, on the face of it, inappropriate to the whole vegetative world."[16] Could a plant will itself to develop seeds? Huxley agreed with Lyell: Lamarck had not provided a plausible mechanism for change; he only gave us "names for things."

A third aspect of Lyell's critique of Lamarck was of particular importance to Huxley's early antitransmutationalist stance. Lyell's uniformitarianism is often contrasted with Cuvier's catastrophism, but when it came to Lamarck, the two men were in agreement. Like Cuvier, Lyell believed that the fossil record provided strong evidence against transformation. Lyell cited approvingly Cuvier's conclusions that the finds from the Napoleonic Egyptian Expedition (1798–1799) showed that the mummies of various animals were the same as present-day species.[17] Furthermore, from the time of the pharaohs, Lyell wrote, domestic animals had been transported by humans to virtually every part of the globe, forced to adapt to every possible climate and habitat. "The cat . . . has been carried over the whole earth . . . from the cold regions of Canada to the tropical plains of Guiana;—yet it has scarcely undergone any perceptible mutation, and is still the same animal which was held sacred by the Egyptians."[18] Surely five thousand years should have been enough time for species to change if they were as pliable as Lamarck had maintained.

Huxley, in his later years, admitted that although he had read *Philosophie zoologique* carefully, it had made little impression on him. "To any biologist whose studies had carried him beyond mere species-mongering in 1850, one half of Lamarck's arguments were obsolete and the other half erroneous or defective, in virtue of omitting to deal with the various classes of evidence which had been brought to light since his time."[19] Furthermore, Lyell's "trenchant and effective criticism" argued not only against transmutation, but also in favor of distinct morphological types.[20] Nevertheless, I want to suggest that Lyell's treatment of Lamarck was not as negative as Huxley claimed. As Michael Bartholomew noted in his article on Lyell's reaction to the possibility of an evolutionary account of man's ancestry, "Lyell's presentation of the species problem, as it stood in 1832 is, in effect, if not intention, far less damning of the idea of transmutation than the rejection of Lamarck might imply."[21]

In spite of Lyell perpetuating Cuvier's misconceptions of

certain aspects of Lamarck's theory, he presented Lamarck's overall argument so well that Herbert Spencer became a Lamarckian as a result of reading the *Principles*: "My reading of Lyell, one of whose chapters was devoted to a refutation of Lamarck's views, had the effect of giving me a decided leaning to them. . . . Lyell's argument produced the opposite effect of that intended."[22] Lyell acknowledged that the theory of transmutation was attractive because it allowed naturalists "to dispense, as far as possible, with the repeated intervention of a First Cause."[23] Although he maintained that species were "real" entities, Lyell admitted that sometimes it was difficult to see the boundaries between species and varieties:

> It is by no means improbable, that, when the series of species of certain genera is very full, they may be bound to differ less widely from each other than do the mere varieties or races of certain species. . . . It is almost necessary, indeed, to suppose that varieties will differ in some cases, more decidedly than some species.[24]

Volume 2 of the *Principles* was certainly giving a mixed message about the nature of species.

William Whewell, in a review of the *Principles*, claimed that Lyell had demonstrated that new species and families "have made their appearance exactly as if they had been placed there, each by an express act of the Creator." Yet he admitted that transmutation "is one of those conjectures easily suggested to the spirit of wide and venturous speculation which these studies almost irresistibly call into action."[25] Indeed, Lyell's critique of Lamarck did not prevent either Wallace or Darwin from developing a theory of transmutation. Lyell's discussion of the distribution, the extinction, and the creation of species, and of competition, and variation, writes Bartholomew, "reads like a scenario for the 1840s and 1850s, with Darwin and Wallace cast in the role of 'the student.' "[26] Although Huxley did not develop his own theory of transmutation, I would suggest that he should be included with the names of Darwin and Wallace. The difference is that Huxley learned a different lesson from his teacher.

Huxley and Lyell— Stepping Out of Darwin's Shadow

Huxley's relationship with Lyell extended beyond reading the older man's book. Huxley first met Lyell in the fall of 1850 while dining at the Geological Club in London.[27] In the spring of 1853, the two men began a correspondence that continued almost until the time of Lyell's death in 1875. The two men soon established a close working relationship. In 1853, obviously impressed with the young Huxley, Lyell sent him the proofs of the eighth edition of the *Principles* and asked Huxley to send him a figure of the larva *Pecten* to use in the book.[28] Throughout the 1850s a somewhat erratic correspondence between the two men continued, mainly about various fossil specimens. This correspondence, viewed in light of Lyell's species journals, shows that they discussed the problem of transmutation extensively.

First of all, it should be remembered that Lyell was a confidant of Darwin almost from the beginning of Darwin's speculations about species change. According to Leonard Wilson, Lyell had known as early as 1837 that Darwin had begun doubting the fixity of species.[29] Huxley, describing the reception of the *Origin*, recalled:

> I remember, in the course of my first interview with Mr. Darwin, expressing my belief in the sharpness of the lines of demarcation between natural groups and in the absence of transitional forms, with all the confidence of youth and imperfect knowledge. I was not aware, at that time, that he had then been many years brooding over the species question; and the humorous smile which accompanied his gentle

answer, that such was not altogether his view, long haunted
and puzzled me.[30]

We have no record of the exact date of this interview, but from
Huxley's further remarks and other correspondence, it appears
that this meeting took place somewhere around 1851–1852.
Huxley, for his part, was interested in developing a classificatory
scheme which would reflect morphological and functional similarities between organisms. Given Huxley's interests and the
fact that Lyell was already aware of some of Darwin's speculations, it was inevitable that Huxley and Lyell would discuss
transmutation. Darwin was the catalyst for these discussions,
but Huxley and Lyell's examination of the problem led them to
disagree with Darwin's gradualistic scheme.

Although there is no mention of transmutation until 1859 in
the exchange of letters between Huxley and Lyell, Lyell's species
notebooks tell a different story. On April 16, 1856, Darwin and
Lyell discussed the species question. Apparently this was the first
time Darwin fully explained his theory of natural selection to
Lyell.[31] Lyell immediately wrote in his journal a summary of the
main features of Darwin's theory.[32] A few days later, he made
another entry under the heading "Origin and Reality of Species"
indicating that he and Huxley had been discussing the matter.

> After conversation with Mill, Huxley, Hooker, Carpenter, &
> Busk at Philos. Club, conclude that the belief in species as
> permanent fixed & invariable, & as comprehending individ-
> uals descending from single pairs or protoplasts is growing
> fainter—no very clear creed to substitute. Indefinite time &
> change may, according to Lamarckian views, work such alter-
> ation as will end in races, which are as fixed . . .[33]

How do species come in to being? Do species multiply from
permanent varieties? Lyell continued:

> The Oxslip, says Huxley, is not known on the continent. It is a
> British permanent variety or race of the Primrose. Would it not
> be preferable to believe in the somewhat sudden development
> of new organs, say the sceptics, than the creation of a new
> species out of inorganic matter or out of nothing? . . . Huxley
> shows that types intermediate between Mammalia and Birds &
> other great divis. are not met with in Geol.ʸ as they w.ᵈ be if
> there was a continual developm.ʳ from one original type.[34]

Lyell's doubts about the origin of species had grown extreme.
He could no longer hold to his older, theologically tinged
belief that species appeared suddenly out of inorganic matter or
out of nothing. Yet, according to Huxley, no evidence indicated
the gradual development of new species either.

From Lyell's letters it is apparent that Darwin had also
taken Huxley into his confidence. On April 30, 1856, Lyell
wrote to Charles Bunbury,

> When Huxley, Hooker, and Wollaston were at Darwin's last
> week, they (all four of them) ran a tilt against species further
> I believe than they are deliberately prepared to go. I cannot
> easily see how they can go so far, and not embrace the whole
> Lamarckian doctrine.[35]

Lyell again mentioned Huxley's discussion of the oxslip. If it
had been found only on Madeira and nowhere else, Lyell
reflected, everyone would have considered it a true species. The
next day Lyell wrote Darwin: "I hear that when you & Hooker
& Huxley & Wollaston got together you made light of all
species & grew more & more unorthodox."[36] As Darwin began
to share his ideas with more people, discussions about transmutation intensified.

Darwin was not the first to describe the difficulty in distinguishing species and varieties. As previously mentioned, Lyell had
discussed the matter in the *Principles* in 1831, and Lamarck had
raised the same issue even earlier. Nevertheless, neither Lyell nor

Huxley had been convinced by Lamarck's theory of species change. However, after grasping the central concepts of Darwin's theory, both men were forced to rethink their objections to transmutation. Right before the publication of the *Origin*, Lyell wrote a long and quite interesting letter to Huxley on the problem of species appearing and disappearing in the fossil record.

> Linnaeus, I think says somewhere that genera really exist. Yet if we find in geology and in nature at present no transition between subkingdoms or even as you hint between class, order and even genera—if there are certain fixed and absolutely limited groups can we explain this according to the theory of transmutation? Must we not suppose that such groups to have come into the world by virtue of some *modus operandi* different from *gradual* development in the course of 1,000 generations? Each newly created type (meaning by creation some act or law not yet within the domain of science) may have been, as a general rule, in advance of the highest of the preexisting ones, Man being the last of the series.[37]

Thus Lyell, the great gradualist in geology, did not see how species could gradually come into being.

Lyell agreed with Linnaeus that genera existed as real entities and mentioned that Huxley had suggested that the same might be true of classes and orders as well. Lyell's letter strongly reaffirmed Huxley's idea of type. But, like Huxley, Lyell was forced to reconsider the idea of transmutation because of Darwin. However, the lack of intermediate forms was problematic. Lyell continued:

> If the Lamarckians are right we shall in time discover extinct fossil varieties of men intermediate between some of the quadramana and Man. If we cannot come to any conclusion in regard to Man the bringing to light of the mode of coming in of antecedent types is hopeless.[38]

How could the absence of transitional forms be made compatible with a theory of transmutation? In his reply to Lyell, Huxley attempted to come to terms with this aspect of Darwin's theory.

> I do not exactly see the force of your argument that we are bound to find fossil forms intermediate between men and monkeys in the Rocks. Crocodiles are the highest reptiles as men are the highest mammals, but we find nothing intermediate between *crocodilia* and *lacertilia* in the whole range of the Mesozoic rocks. How do we know that Man is not a persistent type?[39]

Persistent types were organisms that were found in the fossil record for long periods of time and remained relatively unchanged. Huxley claimed that a great number of them existed. They embodied his idea of distinct morphological types. In this stage of his thinking, Huxley was not particularly interested in finding transitional forms, since they would contradict his deeply held belief in the type concept. The fact that the fossil record did not show evidence of transitional forms between mammals and reptiles or between other major groups suited Huxley perfectly. He did not think it was necessary to find some extinct animal that used primitive tools. It was enough that savages still existed who used bones and twigs. "Why should *Homo eocenus* or *Ooliticus* have been more advanced?"[40] Accepting the racialist attitudes of his time, Huxley believed that savages were clearly inferior to the modern Englishman. Modern-day savages perhaps were essentially the same as *Homo eocenus* and suggested that Man also was a persistent type. The absence of transitional forms did not rule out transmutation for him, but it created problems for one aspect of Darwin's theory. The gaps in the fossil record did contradict Darwin's idea of gradual change, of *natura non facit saltum.*

Saltation as a Solution to Huxley's Dilemma

Huxley was grappling with the concept of type, the lack of transitional forms, and uniformitarianism. How could he put all these ideas together in a way that would be compatible with the theory of transmutation? Huxley's solution to this quandary was saltation. As he explained to Lyell,

> Suppose that external conditions acting on species A give rise to a new species, B; the difference between the two species is a certain definable amount which may be called A-B. Now I know of no evidence to show that the interval between the two species must *necessarily* be bridged over by a series of forms, each of which shall occupy, as it occurs, a fraction of the distance between A and B.[41]

He suggested that the Ancon sheep and the six-fingered Maltese family described by Réaumur were examples of new forms produced all at once with no intermediate stages. Perhaps, Huxley thought, the laws which governed the modification of organic bodies might be similar to chemical laws. Thus, when one atom substituted for another in a chemical compound, each modification was a discrete step; there were no intermediate stages.

> The fixity and definite limitation of species, genera, and larger groups appear to me to be perfectly consistent with the theory of transmutation. In other words, I think *transmutation* may take place without *transition*. . . . *Natura fecit saltum.* . . . The transmutation hypothesis . . . is the logical development of Uniformitarianism and that its adoption would harmonize the spirit of paleontology with that of Physical Geology.[42]

Darwin's theory was a naturalistic explanation of species change, just as uniformitarianism was a naturalistic explanation for geological change. Moreover, chemistry, a discipline that had recently established itself as a science free of metaphysical speculation, demonstrated that saltative evolution need not imply some kind of miraculous intervention. Rather, for Huxley, saltation was a logical extension of Lyell's geological *and* biological views. Saltation allowed Huxley to explain the gaps in the fossil record, accept evolution, and, most important, maintain a belief in the concept of type.

Lyell played a significant role in Huxley's thinking about evolution, but a somewhat different role than the one he played for Darwin. Like Darwin, Huxley was attracted to Lyell's naturalistic explanation of earth history. Uniformitarianism meant that there was no need to bring in the idea of special creation to explain the physical world. But, unlike Darwin, Huxley did not apply the Lyellian idea of slow, gradual change for the biological world. Yet (what has been overlooked in attributing Darwin's gradualism to Lyell's influence) Lyell was *not* a gradualist when it came to the organic world. His view of species change was essentially saltationalist. By some unknown but naturalistic means the earth was successively repopulated with different organisms in response to changing conditions. Lyell did not envision organisms gradually changing to adapt to a changing environment. Rather, new organisms appeared all at once, perfectly adapted. Huxley discussed his saltationalist views with many people, including Darwin, after the publication of the *Origin*. However, the correspondence between Huxley and Lyell, Lyell's treatment of species in the *Principles*, and particularly Lyell's species journals strongly suggest that Lyell was crucial to Huxley's formulation of his saltationalist views.

While Lyell and Huxley were both saltationalists with respect to species origins, the two men differed in important ways in their thinking about transmutation. Lamarckian and Darwinian theories were significantly different; yet Lyell regarded Darwin as essentially a modification of Lamarck. In an earlier quoted passage Lyell writes, "if the Lamarckians are right"; yet this letter was written in regard to Darwin's theory, not Lamarck's. Huxley immediately recognized the difference between the two theories. Although Lamarck gave him no reason to change his "negative and critical attitude" about transmutation,[43] Huxley immediately embraced Darwinism, in spite of the problems the theory had for him.

After the *Origin* appeared, Lyell's opinion of Lamarck improved, and he urged Huxley to reread him. Huxley agreed to, but replied,

> I doubt if I shall improve my estimate of [him]. The notion of common descent was not his—still less that of modification by variation—and he was as far as De Maillet from seeing his way to any *vera causa* by which varieties might be intensified into species. If Darwin is right about natural selection—the discovery of this *vera causa* sets him to my mind in a different region altogether from all his predecessors—and I should no more call his doctrine a modification of Lamarck's than I should call the Newtonian theory of the celestial motions a modification of the Ptolemaic system.[44]

Six months later, Lyell was still comparing Darwin to Lamarck, much to Darwin's consternation:

> You refer repeatedly to my view as a modification of Lamarck's doctrine of development and progression. If this is your deliberate opinion there is nothing to be said, but it does not seem so to me. Plato, Buffon, my grandfather before Lamarck, and others propounded the *obvious* views that if species were not created separately they must have descended from other species, and I can see nothing else in common between the "Origin" and Lamarck. I believe this way of putting the case is very injurious to its acceptance as it implies necessary progression, and closely connects Wallace's and my views with what I consider, after two deliberate readings, as a wretched book, and one from which (I well remember my surprise) I gained nothing.[45]

Darwin's attack on Lamarck seems extreme. Indeed, he was far more influenced by Lamarck than he was willing to admit. However, his letter points out another important issue in the debates over evolutionary theory, and one in which Huxley became thoroughly enmeshed: the problem of progression. It is to this subject that I turn in the following chapter.

Notes

1. Thomas Huxley, 1887, "On the Reception of the *Origin of Species*," *Life and Letters of Charles Darwin* (*LLCD*), 2 vols., Francis Darwin, ed. (London: John Murray, 1887), 1: 55l.

2. Huxley advocated saltation until about 1867–1868.

3. Thomas Huxley, *LLCD* 1: 543–44.

4. A great deal of secondary literature on uniformitarianism exists. A few relevant citings include the following: Peter Bowler, *Evolution* (Berkeley: University of California Press, 1984); Walter Cannon, "The Uniformitarian-Catastrophist Debate," *Isis* 51 (1960): 38–55; Charles Gillispie, *Genesis and Geology: A Study in the the Relations of Scientific Thought, Natural Theology and Social Opinions in Great Britain 1790–1850* (1951; New York: Harper, 1959); Stephen Gould, "The Eternal Metaphors of Paleontology," in *Patterns of Evolution*, Anthony Hallam, ed. (Amsterdam: Elsevier, 1977), pp. 1–26; Stephen Gould, "Agassiz's Marginalia in Lyell's *Principles* or the Perils of Uniformity and the Ambiguity of Heros," *Studies in the History of Biology* 3 (1979): 119–38; Reijer Hooykaas, *Natural Law and Divine Miracle; the Principle of Uniformity, Biology, and History* (Leiden: Brill, 1959); and Martin Rudwick, "Uniformity and Progression: Reflections on the Structure of Geological Theory in the Age of Lyell," in *Perspectives in the History of Science and Technology*, Duane H. Roller, ed. (Norman: University of Oklahoma Press, 1971), pp. 209–27. I find Hooykaas's presentation on the distinction between uniformitariansim and catastrophism particularly clear. As will be developed in the next chapter, I agree with Cannon that after the *Principles* appeared, the controversy in geology over evolutionary theory was not between catastrophism and uniformitarianism. Rather, debate centered around the problem of progressive development.

5. Important exceptions include William Coleman, "Lyell and the 'Reality' of Species: 1830–1833," *Isis* 53 (1962): 325–38; Leonard Wilson, ed., *Sir Charles Lyell's Scientific Journals on the Species Question* (New Haven: Yale

University Press, 1970); Michael Bartholomew, "Lyell and Evolution: An Account of Lyell's Response to the Prospect of an Evolutionary Ancestry for Man," *British Journal for the History of Science* 6 (1973): 261–303; and Michael Bartholomew, "The Singularity of Lyell," *History of Science* 17 (1979): 276–93.

 6. See Wilson, *Sir Charles Lyell's Scientific Journals.*

 7. Charles Lyell, 1830–1833, *The Principles of Geology,* 3 vols. (London: John Murray; Chicago: University of Chicago Press, 1991), 2: 1.

 8. Ibid., Lyell quoting Lamarck in *Philosophie zoologique,* tom.i. p. 54, *Principles,* p. 3.

 9. Ibid., p. 4.

 10. Ibid., p. 7.

 11. Ibid., p. 8.

 12. Ibid.

 13. Thomas Huxley quoting Lamarck, "The Darwinian Hypothesis," *Collected Essays of Thomas Huxley,* vol. 2, *Darwiniana* (London: Macmillan & Co., 1893), p. 12.

 14. Ibid., pp. 12–13.

 15. Lyell, *Principles,* 2: 9–10.

 16. Huxley, 1887, "On the Reception of the *Origin of Species,*" *LLCD* 1: 543.

 17. Naturalists accompanying the Expedition had sent to Paris large numbers of mummified specimens collected from the tombs and burial grounds along the Nile. Cuvier had examined these specimens (Coleman, "Lyell and the 'Reality' of Species"), Lyell, *Principles,* 2: 29–30.

 18. Lyell, *Principles,* 2: 30.

 19. Huxley, "On the Reception of the *Origin of Species,*" *LLCD* 1: 543.

 20. Ibid.

 21. Bartholomew, "Lyell and Evolution," p. 284.

 22. Herbert Spencer, *An Autobiography,* 2 vols. (New York: D. Appleton & Co., 1904), 1: 201.

 23. Lyell, *Principles,* 2: 18.

 24. Ibid., p. 23.

 25. William Whewell, "Principles of Geology . . . , vol. ii (1832)," *Quarterly Review* 47 (1832): 109.

 26. Bartholomew, "Lyell and Evolution," p. 284.

 27. Huxley to his sister Lizzie, November 21, 1850, *LLTHH* 1: 68.

 28. Lyell, April 1853, Huxley Manuscripts (HM) 6: 1.

 29. Wilson, *Sir Charles Lyell's Scientific Journals,* p. xlv.

 30. Huxley, "On the Reception of the *Origin of Species*" (*LLCD*) 1: 549–50.

31. Wilson, *Sir Charles Lyell's Scientific Journals*, pp. xliv–xlv. See also H. Lewis McKinney, "Alfred Russel Wallace and the Discovery of Natural Selection," *Journal of the History of Medicine* 21 (1966): 333–57.

32. Charles Lyell, April 16, 1856, *Scientific Journal* 1, in Wilson, *Sir Charles Lyell's Scientific Journals*, pp. 54–55.

33. Ibid., pp. 56–57.

34. Ibid., p. 57.

35. Charles Lyell, April 30, 1856, *Life, Letters and Journals of Sir Charles Lyell* (*LLCL*), 2 vols., Katherine Lyell, ed. (London: John Murray, 1881), 2: 212.

36. Charles Lyell, May 1, 1856, Darwin mss., quoted in Wilson, *Sir Charles Lyell's Scientific Journals*, p. xlvii.

37. Charles Lyell, June 17, 1859, HM 6: 120.

38. Ibid.

39. Thomas Huxley, June 25, 1859, *LLTHH* 1: 187.

40. Ibid.

41. Ibid., p. 185.

42. Ibid., pp. 185–87.

43. Huxley, "On the Reception of the *Origin of Species*," *LLCD* 1: 543.

44. Huxley to Lyell, August 17, 1862, *LLTHH* 1: 215.

45. Darwin to Lyell, March 12, 1863, *LLCD* 2: 198–99.

Chapter Four

Fossils, Persistence, and the Argument from Design

Gradualism was not the only aspect of Darwin's theory that Huxley found problematic. In addition, Huxley argued against progression in the fossil record, which caused him difficulties in trying to simultaneously support Darwin as well. Huxley's views on progression were the result of a complex interaction of a variety of factors. Ironically, his antiprogressionist stance was due in large part to Charles Lyell, the font of much of Darwin's own inspiration. In his famous work of the 1830s, *The Principles of Geology*, Lyell had claimed that the fossil record documented no directional change. Indeed, some types seemed to persist unchanged through vast periods of geological time. While a strong case could be made against progression as late as the 1850s, the fossil discoveries in the next decade began more forcefully to evince progressive development. Mammals simply were not turning up in the Silurian epoch. Yet Huxley remained fixedly disposed against the idea of progression. I suggest that Huxley had another reason for maintaining a belief in nonprogression: In his mind progression was linked to the argument from design.

Huxley's desire to keep theological questions distinct from scientific ones underlies virtually all of his scientific work. As

Jacob Gruber wrote, "He wanted to clear the garden of science from the weeds of philosophy, pseudo-philosophy, and theology which had invaded it."[1] For the most part, such an attitude served Huxley well. However, while Huxley's objections to various aspects of Darwin's theory were firmly grounded in what he believed was lack of adequate empirical evidence, the reasons for his denial of progression were much less straightforward. I believe that his antitheological fervor resulted in him denying progression far longer than the evidence warranted. Essential to my analysis is situating Huxley's views within contemporary debates in geology over fossil progression.

A Directionalist View of Earth History

Chapter 3 mentioned three aspects of Lyell's uniformitarianism: The history of the earth could be explained in terms of causes that were naturalistic, actualistic, and gradual. However, there was a fourth component to Lyell's theory, which revived James Hutton's view that the history of the earth exhibited a steady-state pattern of endlessly repetitive cycles leading nowhere in particular. As Hutton wrote, "We find no vestige of a beginning—no prospect of an end."[2] In contrast, catastrophists viewed earth history developmentally and maintained that general overall trends through time could be detected on a global scale. This directional view of earth history was based on the geophysical theory of a cooling earth. Buffon had popularized the idea of a gradually cooling earth in his speculative Époques de la Nature (1778). By the 1820s the theory enjoyed considerable prestige. First of all, the idea of a "central heat" had considerable empirical support. Louis Cordier had shown that a geothermal gradient was not only real, but universal. In addition, the application of Fourier's physics had been used to explain both the idea of a cooling earth through time and the geothermal gradient. Fourier had demonstrated that the temperature along the earth's surface plotted against time had fallen along an exponential curve. In the recent epochs the heat loss from the earth's interior was quite small compared to the effects of solar radiation, resulting in fairly uniform and stable conditions.[3] The evidence for directional change continued to mount, and Lyell's steady-state view of geological change made few converts.

A theory of residual central heat implied a directionalist interpretation for a variety of phenomena. For example, if the only adequate source of volcanic heat was the earth's central heat, vulcanism would decrease over time. Thus, the theory of a cooling earth implied that environmental conditions on the earth were very different in the ancient epochs than in more recent ones. It was not unreasonable to conclude that such different physical conditions meant the organic world would also be vastly different. At first the world was too hot for any kind of life. Reptiles dominated the secondary period because they were suited to a tropical environment. As the earth continued to cool, mammals made their appearance in the Tertiary. The many new fossil discoveries were providing evidence for just such a pattern.

Cuvier's work clearly demonstrated that organisms had become extinct and entirely new faunas had replaced them. As the environment, particularly climatic conditions, changed, organisms had to change in response. By "changed," I do not mean transmuted, but rather that new organisms appeared that were adapted to the particular environment in which they were found. They flourished for a time, but as conditions changed, they became extinct and were replaced with other organisms that were better adapted to the new environment. Thus, a developmental view of earth history was based not just on geological change, but on changes in the organic world as well. As Rudwick has pointed out, organic and geological change were linked, each providing support for the other. A directionally changing environment could explain the directionally changing nature of successive faunas.[4] But did this mean the change was progressive? An organism might be different, but not necessarily more complex. Nevertheless, most researchers who argued for the successive appearance of different organisms also believed that the changes were progressive, that is, that higher and more elaborate forms came into existence.

By the 1830s the idea of progressive development was generally well accepted among geologists. What later became known as stratigraphical geology demonstrated that the earth was at least several hundred thousand years old, with organisms of increasing complexity appearing over time. The discovery of many fossils of extinct organisms in the eighteenth century had already brought into serious question the creation myth as a complete explanation of the earth's history. Most geologists rejected the limited time scale of the scriptural geologists who claimed that the fossil record could be compressed into the restricted amount of time a literal interpretation of the Bible would require.[5] This did not mean they were unconcerned with reconciling geology and Christianity. Natural theology as propounded by William Paley and the Bridgewater Treatises (which will be detailed shortly) sought to integrate the new geological findings with traditional Christian teaching. Some catastrophists, such as James Parkinson, William Buckland, and Louis Agassiz, argued that the changing conditions of the earth reflected God's plan to gradually prepare for the coming of humanity. Nevertheless, by 1830 few professional geologists believed that God was continually performing miracles. Rather, God's presence was manifested in the original design of the system. The argument from design claimed that the order and complexity of the world, especially the adaptation of each individual to its environment, could only be the result of an intelligent designer. Natural theology thus gave support to directionalism, which in turn fostered a progressive interpretation of the fossil record.[6] But actualism combined with a steady-state view of the world didn't *lead* Lyell to deny progression—it *is* a denial of progression. Since species were adapted to their environmental conditions, which were cyclic and recurrent, how could there be progression? However, the story is more complicated than this, since one could still advocate a progressive actualism, as did Darwin.[7] As will be shown, Huxley and Lyell denied progression because of their particular philosophical and theological worldviews.

The Fossil Evidence

When Lyell wrote the *Principles of Geology*, paleontology as a rigorous discipline was barely twenty years old. The accumulated findings yielded an extremely incomplete record, and thus interpretation was based on a very fragmented pattern. Lyell argued that "reading" the fossil record was like reading a book that had most of the pages missing and few words left on the remaining pages. Darwin was impressed with Lyell's analogy and used it to explain the gaps in the fossil record—the absence of transitional organisms. He said to a friend, "It strikes me that all of our knowledge about the structure of the earth is very much like what an old hen would know of a hundred acre field, in a corner of which she is scratching."[8] For Darwin, evolution was continuous and gradual, and Lyell's analogy provided an explanation for why the fossil record shouldn't be used as evidence against his gradualistic scheme. But this was not the original intent of Lyell's argument. Rather, the purpose of the analogy was to argue against progressive development. The only reason the fossil record looked progressive, he concluded, was because of its incompleteness.[9] Since sampling of the record was quite limited and displayed many anomalies, the case for progression was far from conclusive.

Two distinct problems confronted the supporters of progressive development. First, as more and more fossils were discovered, the origin of each vertebrate class was pushed further and further back in time. Lyell used these findings to argue that all classes were present at all times in geological history. He claimed that forms that appeared late in the fossil record, such as mammals, eventually would be discovered in the early epochs. Second, and more problematic, often the earliest known examples of a class exhibited a more complex level of organization than later occurring members. If the fossil record did not show progression *within* groups, how could one claim that overall it was yet progressive? Supporters of discontinuous progression countered that it didn't matter whether progressive development occurred within groups or not. They maintained that the fossil record clearly showed that groups of higher and more complex organization appeared later in time. The record in the rocks indicated that invertebrates appeared in the Silurian, the bony fishes in the Devonian, the reptiles after the Carboniferous, and the birds and mammals in the Tertiary period. Nevertheless, Lyell and Huxley were able to point out enough inconsistencies in the fossil record to maintain a lively debate in the geological community.

Much of the debate between progressionists and nonprogressionists was carried on at the meetings of the Geological Society of London, which in the mid-1800s was one of the most prestigious scientific societies in Europe. The annual addresses of the president dealt with the latest geological findings and discussed areas of controversy. Each president could use these lectures as a forum for his particular views. Both Lyell and Huxley used their presidential addresses to argue against progression.

In his 1851 anniversary address to the society, Lyell summed up the evidence against progression.[10] Huxley would use many of the same examples to argue for persistent types and build his own case against progression. The underlying theme of the address was that the fossil record was far too fragmentary to make any sweeping generalizations about the nature of earth history. Since particular fossils represented only the community of flora and fauna found at a specific locality, they could not represent the entire life of the earth's surface at the time they were deposited. The oldest strata known were all marine formations. That did not mean, however, that land masses did not exist in the Silurian period, only that we would have no knowledge of them. The most dramatic evidence in support of this thesis was the then-recent discovery that sediments in modern seas poorly reflected the diversity in the contemporary living world. In 1850, Edward Forbes and Robert MacAndrew had made a series of dredgings off the coast of Great Britain. While they brought up fossils of many marine invertebrates and fish, they found no remains of Cetacea or land mammals.[11] Although the discoveries of Forbes and MacAndrew were important, Lyell relied mainly on fossils from ancient strata "to establish [his] opposition to the theory of successive development."[12]

Lyell told the audience for his 1851 address that he had twelve principle points which he expected "to establish in opposition to the theory of successive development."[13] I will spare the reader a discussion of all twelve points, but a few comments are in order. In 1849, the continental paleobiologist Adolphe-Théodore Brongniart had argued for the progressive nature of the plant world in his *Tableaux des Genres des Vegetaux Fossiles*, and it was accepted as the most authoritative treatise on fossil plants. However, Lyell cited several examples from the plant world to make two main points. First, the earliest fossil plants did not represent the simplest level of organization. All classes of plants existed from the Cretaceous through the Tertiary period.[14] Although a complete turnover in species had occurred four or five times, no significant advance in plant organization or complexity had occurred. Second, these fossils came from marine deposits and therefore could not give us information about the terrestrial vegetation of the same period. However, even with the fragmentary nature of the fossil record, highly complex plants had been found from some of these earliest periods.[15]

Lyell then turned his attention to the animal world. The oldest Silurian strata had highly developed representatives of the three main classes of invertebrates. Lyell also mentioned Agassiz's important work on fossil fish. Although Agassiz believed in the successive appearance of the vertebrate classes, he admitted that his work on fish demonstrated no increase of organization within the class through time. Furthermore, saurians in the Permian had as high a level of organization as any existing today.[16]

Several recent finds helped Lyell's antiprogressionist position. Recently, skeletons of highly developed reptiles as well as reptilian footprints had been discovered from the Carboniferous epoch. William Logan had found in the Silurian rocks of Canada some footprints that Owen claimed were made by a freshwater tortoise.[17] Soon afterward, Gideon Mantell found the remains of a reptile he named *Telerpeton elginense* in what was thought to be the Old Red Sandstone.[18] Both these discoveries put the origin of reptiles far earlier than anyone but Lyell or Huxley could have imagined. Edward Hitchcock claimed he had found bird tracks in the New Red Sandstone of the Connecticut valley. Although these footprints were later identified as dinosaur tracks, at the time the consensus was that they had been made by birds.[19] In addition, Lyell mentioned that Owen was now beginning to think that the Oolitic "marsupials" might actually be true placentals.[20] If Owen was right, then mammals which were generally regarded as more highly developed than marsupials were present much earlier in the fossil record than anyone had thought possible. Summarizing the mammalian fossil record, Lyell admitted that mammalian species had changed five or more times in the Tertiary period. Yet he still claimed that remains from this period indicated that the species were as highly organized as those existing today.[21]

Besides citing various fossil finds to support his antiprogressionist cause, Lyell could also quote a few well-respected paleontologists sympathetic to his views. While Gideon Mantell had made many discoveries toward establishing the Age of Reptiles, he still accepted Lyell's antiprogressionism. In the *Wonders of*

Geology, he pointed out that the Stonesfield mammals existed in the middle of the Age of Reptiles and that "some of the fossil animals which first appear in the strata belong to families with a highly developed organization."[22] Edward Forbes believed his own research indicated that "the scale of the first appearance of groups of any degree is most clearly not a progressive one."[23] Lyell also had support among some continental paleontologists. In 1845, Constance Prévost argued that there was no evidence that living forms had become more perfect over time either in response to changing conditions or as a result of the unfolding of a hierarchical divine plan.[24] A. D. D'Orbigny admitted that the order of appearance of vertebrate classes was progressive, but within classes the evidence was contradictory. Furthermore, he claimed that if the general trend was progressive, then primitive orders should decline in the number of species, while more advanced orders should show a corresponding increase. However, the fossil record indicated that the number of primitive species had actually increased over time. Like Lyell, D'Orbigny believed that any increase in the level of organization was an illusion.[25] Lyell noted that both Prévost and D'Orbigny were "not satisfied with the paleontological evidence in support of the doctrine of successive development."[26] While these paleontologists were sympathetic to Lyell's views, it was Huxley who actively promulgated the nonprogressionist cause. As he was later to do for the theory of evolution, Huxley mixed detailed scientific evidence and biting sarcasm in his often highly polemical defenses of nonprogression.

In a scathing review of the tenth edition of the anonymously published *The Vestiges of the Natural History of Creation*,[27] Huxley asked, "Has the progression theory any real foundation in the facts of paleontology? We believe it has *none*."[28] Reiterating much of the evidence that Lyell had used in his 1851 address as well as adding other examples, Huxley devoted several pages to documenting the lack of progression in the fossil record. Ferns, angiosperms, and various other higher plants appeared in the Devonian and even the Silurian periods. Some groups of plants were more highly organized in the Carboniferous epoch than their present-day representatives. Similar arguments were made for animals. Trilobites appeared earlier than annelids, although the trilobites were more complex.

Huxley pointed out, as had many other critics, that the Vestigiarian (the author of the *Vestiges*) distorted both the fossil record and accepted classificatory schemes in order to make the record appear progressive. For example, sponges and foraminifera were easily preserved because of their calcareous spicula and shells. Yet barely a trace of them had been found in the lowest strata. Standard classifications considered them less developed than the polypian family, and therefore they should appear earlier in the fossil record. But the Vestigiarian treating "the forms of animals in ascending order, 'illustrates' his own geological lore by placing foraminifera [*sic*] *after* polypiaria."[29]

Huxley discussed in great detail the Paleozoic fishes. The Vestigiarian cited Agassiz, who claimed that the Ganoid and Placoid fishes from the Devonian represented the embryonic stages of osseous or "more perfect fishes." But Huxley believed that Agassiz's "lively fancy has a done at least as much harm to natural science as his genius has assisted its progress."[30] Although Huxley railed against Owen's platonic concept of types, he did not hesitate to cite Owen when it suited his own purposes. Referring to Owen's Hunterian Lectures as an "excellent work," Huxley quoted from them extensively to discredit the Vestigiarian's classification of fish based on the amount of cartilage in the vertebra and the incompleteness of the ossifying process.[31] According to Owen, many of the most highly organized fish appeared early in the fossil record. Therefore, the amount of ossification was not a good means for determining the level of organization. Although Owen did think that overall the fossil record progressed, his belief in archetypes meant that organisms should be "rated" for perfection within their own type rather than looking in their structure for indications of a different or higher type. Huxley pointed out that all of chapter 6 of the Hunterian Lectures was "devoted to a most successful demonstration of the non-embryonic nature of cartilaginous fishes, and the author speaks not without some contempt of the progressionists."[32]

> Yet there are some who would shut out, by easily compre-
> hended quite gratuitous systems of progressive transmutation
> and self-creative forces, the soul-expanding appreciations of
> the final purposes of the fecund varieties of the animal struc-
> tures, by which we are drawn nearer to the great First Cause.
> They see nothing more in this modification of the skeleton,
> which is so beautifully adapted to the exigencies of the *highest
> organized of fishes*, than a foreshadowing of the cartilaginous
> conditions of the reptilian embryo in an enormous tadpole,
> arrested at an incomplete stage of typical development. But
> they have been deceived by the common name given to the
> plagiostomous fishes: the animal base of the shark's skeleton
> is not cartilage.[33]

Citing Owen's Hunterian Lectures undoubtedly served another purpose of the young Huxley—to discredit Owen. How devilishly clever. This would not be the only time that Huxley would use Owen's earlier writings to undermine his present position.

In addition to Agassiz, the Vestigiarian had cited an anonymous article in the *Quarterly Review* about Lyell's work, in support of progression, which he credited to Owen.[34] Huxley also agreed that the article was by Owen, writing to William MacLeay "that the review has done him much harm in the estimation of thinking men—and curiously enough, since it was written, reptiles have been found in the old red sandstone, and insectivorous mammals in the Trias!"[35] While citing Owen favorably in many cases, the article also quoted a passage from his *Comparative Anatomy*, but the reviewer claimed the quotation represented "a statement hazarded by the advocate of a particular view—not the generalization which the equal ponderer on *all the phenomena* would have enunciated."[36] However, Huxley had not depicted the *Quarterly Review* article accurately. Quoting only selective parts, he does not do justice to the complexity of Owen's argument. The review was primarily a critique of Lyell's 1851 address. The reviewer argued that the fossil record indicated that over time, the development of various groups had resulted in an "exchange of a more general for a more specialized type. The modifications which constitute the departure from the generalized type adapts the creature to special action and usually confers upon it special powers."[37] One example he used was the development of the horse, which ironically was the same example that Huxley would later use in his American addresses as demonstrative proof of evolution. The reviewer claimed that invertebrates preceded vertebrates and cold-blooded organisms preceded warm-blooded ones, but he did not think that Lyell would disagree with this claim. He pointed out that nothing in the writings of Owen or even Miller "opposes the idea that vertebrate organization was coeval with the molluscous, articulate and radiate types on this planet."[38] Thus, Owen was advocating limited progression within general types—a view that was surprisingly similar to one that Darwin would espouse. But Huxley's distortion made Owen appear confused and self-contradictory on the issue of progression. By pointing out Owen's supposed inconsistency in addition to attacking Agassiz's work on fossil fish, Huxley hoped to demolish the progressionist proposition.

Throughout the 1850s, Huxley and Lyell continued to build a case against progression. In 1858, Huxley gave an important paper on the crocodilelike fossil *Stagonolepis*. By comparing it to a crocodile, he gave the impression that crocodiles were living fossils, a group that had remained virtually unchanged since the Triassic.[39] Herbert Spencer wrote to Joseph Hooker after hearing Huxley's talk that the "evening was a triumph for Huxley, and rather damaging for the progressive theory, as [it is] commonly held."[40] Nevertheless, nonprogression was becoming increasingly difficult to defend, and Darwin's theory was a fresh assault on it.

Huxley's Persistent Types

Both Lyell and Huxley had been attacking progression, but neither of them had much use for the idea of transmutation either. Lyell, until the time of the publication of the *Origin*, was, according to Huxley, "a pillar of the antitransmutationists."[41] Furthermore, Huxley claimed that "the evidence in favor of transmutation was wholly insufficient."[42] Yet because of Darwin's theory, both men were forced to reconsider transmutation. As I have said, Huxley adopted saltation as a way to reconcile his belief in type with the idea of transmutation, but this still left the problem of progression. He admitted that most paleontologists claimed that the fossil record documented a "vast contrast between the ancient and the modern organic worlds."[43] But Huxley disagreed.

> Without at all denying the considerable positive differences which really exist between the ancient and modern forms of life . . . an impartial examination of the facts revealed by paleontology seems to show that these differences and contrast have been greatly exaggerated. Thus, of some two hundred known orders of plants, not one is exclusively fossil. Among animals there is not a single totally extinct class, and of the orders, at the outside not more than seven percent are unrepresented in the existing creation.[44]

Among plants, many ferns, club mosses, and *conifera* had not changed since the carboniferous period. In the animal kingdom, certain foraminifera, the squid, and the crocodile seemed to have existed virtually unchanged for vast lapses of time. Such organisms Huxley called persistent types.

Persistent types served a variety of purposes for Huxley. They were, of course, evidence for Lyellian nonprogression. Huxley's own research along with the work of Cuvier and von Baer provided strong evidence for the belief in distinct morphological types. Persistent types allowed Huxley to maintain his belief in type, but they also addressed what Huxley regarded as a significant problem in Darwin's theory and one of which Darwin was also aware. If evolution was progressive, why was progress not universal? As Darwin wrote to Asa Gray soon after the publication of the *Origin*, "Judging from letters . . . and from remarks, the most serious omission in my book was not explaining how it is as I believe that all forms do not necessarily advance, how there can now be simple organisms existing."[45] Huxley asserted that the existence of persistent types was inexplicable except on evolutionary grounds. Darwin's theory was based upon two factors: the tendency to vary and the influence of the environment on the form of those variations. Variation occurred spontaneously, but natural selection organized and selected those variations or adaptations that made an organism well suited for its particular environment. If a parental form was better adapted to the environment, it would persist, and the derived form would perish. If the derived form was better adapted, then the parental one would die out. In the first case there would be no progression, but in the second there would be change from one type to another. Even then the change might not be described as progressive. Furthermore, Darwin also had an account of retrogression, though only hinted at in the *Origin*.[46] Other theories about the history of life claimed progression was inevitable. The fossil record represented the unfolding of God's plan, which was ever progressing, preparing the way for the coming of Man. I shall return to this point later, as my claim that Huxley continues in the 1860s to associate progression with the argument from design in part rests on it.

In his 1862 presidential address to the Geological Society, Huxley reiterated and expanded his 1859 views on persistent types. He argued that the common doctrines of progressive modification which supposes that modification to have taken place by a necessary progression from more to less embryonic forms, from more to less generalised types, is negated by an impartial survey of the positively ascertained truths of paleontology.[47]

Lyell was thrilled by Huxley's lecture, describing it as a "brilliant critical discourse on paleontology, [showing] how much the progressive development system has been pushed too far, how little can be said in favor of Owen's more generalized types ... the persistence of many forms high and low."[48] However, Darwin was not so enthusiastic. Huxley anticipated that people might misinterpret his lecture in relation to Darwin's theory, and he wrote the following to Darwin:

> I want you to chuckle with me over the notion I find a great many people entertain—that the address is dead against your views. The fact being, as they will by and by wake up [to] see that yours is the only hypothesis which is not negated by the facts,—one of its great merits being that it allows not only of indefinite standing still, but of indefinite retrogression.[49]

But Darwin was not chuckling. Huxley's presentation seemed quite one-sided, with all the evidence being cited in favor of persistence and none in favor of progression. Darwin wrote back:

> I cannot help hoping that you are not quite as right as you seem to be. Finally, I cannot tell why, but when I finished your Address I felt convinced that many would infer that you were dead against change of species, but I clearly saw that you were not.[50]

Darwin would have liked a little less persistence and a little more progression.

Huxley's concept of persistent types *was* problematical for Darwin's theory. On the one hand, persistent types could be

used to deal with those organisms that had remained unchanged for millions upon millions of years, yet overall organisms did come into being and disappear. Why did some organisms persist while most did not? In this respect, Darwin's theory
had more in common with the catastrophist belief in directional
change than the steady-state view of Hutton and Lyell. While
Cuvier did not believe in transmutation, he and other catastrophists argued that the fossil record was the best proof that
the history of the earth showed progressive changes. Why
didn't Darwin adopt their views? Certainly the association of
catastrophism with special creation in Great Britain was a factor,
but one cannot underestimate the importance of Darwin's relationship with Lyell. Lyell was not only a scientific colleague,
but a close personal friend as well. How could Darwin adopt
the views of Cuvier in direct opposition to Lyell? Not only did
Darwin share the catastrophist belief in directional change, but
as mentioned, Darwin's paleontology was more like Owen's
than Lyell's, with a genealogical family tree superimposed over
the idea of progressive divergence.[51] This put Huxley in an extremely difficult position. He was a close friend and colleague
of both Lyell and Darwin, and he despised Owen. As much as
he liked Darwin's theory, he was not about to adopt Owen's
progressionism.

Huxley's views on evolution changed in the decade following the publication of the *Origin*. Most notably, he abandoned saltation in favor of gradualism.[52] In 1870, he reassessed
his views on paleontology. Some of his positions he modified,
but progression was not one of them. "The significance of persistent types, and of the small amount of change which has
taken place even in those forms which can be shown to have
been modified, becomes greater and greater in my eyes, the
longer I occupy myself with the biology of the past."[53]
Throughout the 1860s many Carboniferous amphibians had
been found that were as highly organized as their Triassic rela-
tives.[54] In particular, eight or ten distinct genera of labyrintho-

donts had been discovered from the Carboniferous that were "more extensive and diversified than those of the Trias."[55] These amphibians became one of Huxley's prime examples of persistence, showing "that a comparatively highly organized vertebrate type, such as that of the Labyrinthodonts, is capable of persisting, with no considerable change, throughout the period represented by the vast deposits which constitute the Carboniferous, the Permian, and the Triassic formation."[56]

Certainly persistent types existed, but to continue to deny the overall progressive nature of the fossil record was quite an untenable position by 1870. Many of the finds that Lyell had cited as evidence for nonprogression in his 1851 address had turned out to be misidentified or misdated. For example, Owen decided that the tracks in Canada were not made by a tortoise, but by an invertebrate, probably *Limulus*.[57] Furthermore, how could Huxley explain that sediments from the Silurian had preserved nothing higher than mollusks and a few fish, while birds and mammals did not show up until the Mesozoic? Yet Huxley told his fellow paleontologists,

> I confess it is as possible for me to believe in the direct creation of each separate form as to adopt the supposition that mammals, birds, and reptiles had no existence before the Triassic epoch. Conceive that Australia was peopled by kangaroos and emus springing up ready-made from her soil and you will have performed a feat of imagination not greater than that requisite for the supposition that the marsupials and great birds of the Trias had no Paleozoic ancestors belonging to the same classes as themselves. The course of the world's history before the Trias must have been strangely different from that which it has taken since if some of us did not live to see the fossil remains of a Silurian mammal.[58]

Even Lyell had reluctantly accepted a limited amount of progression by 1863 when he published *The Antiquity of Man*.

Huxley, if pushed, admitted that some organisms did progress over vast periods of time. But why did he continue to emphasize persistence when the fossil evidence clearly documented a progressive trend as well? In spite of Huxley priding himself on his empiricism, the answer will not be found just in the rocks.

The Theological Implications of Progression

In the beginning of the chapter, I suggested that progression was associated with the argument from design.[59] In the eighteenth century, the discovery of large numbers of fossils was quite problematic for those committed to an orthodox view of nature. Natural theology attempted to stave off the inevitable conflict between science and religion. Works such as Charles Bonnet's *Contemplation de la Nature* (1764) and Jean Baptiste Robinet's *De la Nature* (1761–1766) claimed that the chain of being represented God's plan of development through time.[60] Underlying Linnaeus's system of classification was the idea that the orderly system of nature was evidence of divine plan.[61] William Paley's *Natural Theology* (1802) continued to espouse the argument from design and was quite influential on nineteenth-century naturalists, including the young Darwin.

In Britain, the influence of natural theology resulted in catastrophism being closely associated with the creation myth. Geological indications that a great deluge had swept across the entire isle was cited as evidence for the biblical flood as told in the story of Noah. The leading proponent of this "scriptural geology" was William Buckland, a Reader in geology at Oxford. In the dedication of the *Reliquiae Diluvianae*, Buckland claimed that he undertook this work to provide evidence such that it could "no longer be asserted that geology supplies no proofs of an event, the reality of which the truth of the mosaic records is so materially involved."[62] Summarizing his evidence, he quoted Greenough's *First Principles of Geology*, which "postulated a succession of upheavals culminating in a "general flood which swept away the quadrupeds from the continents, tore up the solid strata and reduced the surface to a state of ruin."[63] Buckland believed these upheavals were due to the "direct agency of creative interference."[64] This "Diluvialist Geology" was soon discredited, as evidence from various parts of the world did not support such a universal deluge. Furthermore, the fossil record did not correspond to the introduction of species as described in Genesis. However, even if miraculous forces were not involved, most geologists believed that the history of the earth was determined by divine providence. Buckland later admitted that he overestimated the extent of the Deluge; but he remained one of the staunchest advocates of natural theology, gathering evidence which was consistent with a biblical interpretation, even if not proof of specific events.

The eighth earl of Bridgewater had commissioned a series of treatises to be written in support of Paley's *Natural Theology*. Buckland was asked to write one on geology, which provided him with the ideal opportunity to show how each stage in the development of life corresponded to the "varied circumstances and conditions of the earth's progressive stages of advancement."[65] Each organism was perfectly adapted to its environment, demonstrating the Creator's benevolence. As conditions on the earth changed, a series of more complex organisms appeared adapted to those conditions:

> [T]he creatures from which all these fossils are derived were constructed with a view to the varying conditions of the surface of the earth, and to its gradually increasing capabilities of sustaining more complex forms of organic life, advancing with successive stages of perfection.[66]

The successive new forms of life were "a distinct manifestation of creative power transcending the operation of known laws of nature, and it appears to us that Geology has thus lighted a new lamp along the path of Natural Theology."[67] Geology provided such overwhelming evidence "of the Being and Attributes of God," it could only be regarded as "the efficient auxiliary and the Handmaid of religion."[68] But it was precisely this sentiment that both Lyell and Huxley were reacting against.

Peter Bowler suggests that the crucial element of the catastrophist position was the idea of a directional trend, not progression.[69] He claims that the progressive nature of the fossil record was essentially evidence for directionalism, and the "possibility of a specifically progressive trend toward man in the sequence of creations was not an integral part of the catastrophist position."[70] Bowler's interpretation is certainly questionable since leading catastrophists such as Buckland and Sedgwick explicitly argued that the fossil record was evidence of a Divine Plan and that the earth was progressively changing in preparation for the eventual coming of humanity. However, for my purposes, what is important is that in defending his steady-state view against the directionalism of catastrophism, Lyell linked progression with directionalism. Furthermore, both he and Huxley associated progression with the argument from design, and both men wanted questions of theology kept distinct from questions of natural history.

Lyell's *Principles of Geology* was an attempt to separate geology and natural history from theology—an extremely attractive idea to Huxley. Lyell wanted a scientific geology and believed that the catastrophist methodology meant that geology "could never rise to the level of an exact science."[71] Huxley was impressed with the *Principles*, primarily because of the methodology implicit in the concept of actualism. Actualism meant that no supernatural forces need be invoked to explain the history of the earth. But Huxley differed fundamentally with Lyell about the implications of progressive development. Lyell denied progression to preserve his religious views regarding the dignity of Man. Huxley, on the other hand, denied progression in order to attack the religious view embodied in the argument from design.

Lyell and the Dignity of Man

Lyell's 1851 address presented an impressive body of fossil evidence to support his antiprogressionist position. But what was the underlying purpose of the address? Michael Bartholomew has claimed that not only was progression contrary to Lyell's steady-state view of earth history, but it also threatened the privileged status of humanity within a totally naturalistic explanation of earth history.[72] Evidence for such an interpretation can be found not only in the address, but much earlier in the *Principles*.

In the 1820s, Lyell, like most natural philosophers of the time, accepted the idea that the history of life followed a general pattern of increasing complexity and differentiation.[73] In 1827, Lyell "devoured" Lamarck's *Philosophie Zoologique* "with pleasure."[74] However, he noted that Lamarck's argument, "if pushed as far as it must go, if worth anything, would prove that men may have come from the Ourang-Outang."[75] Lyell effectively criticized Lamarck's theory of transmutation in the *Principles*. But while previous progressive interpretations of the fossil record had relied on special creations to explain the history of life, Lamarck's theory, even if wrong, was still a *naturalistic* explanation for progression. Lyell realized that a transmutationist interpretation of progression would link man to beast and that "some sort of evolution would provide the *only possible* naturalistic explanation of species origination, if there had indeed been a progressive sequence of animals and plants through time."[76] Lyell was appalled at the idea that humans might have evolved from a lower form. Thus, he balked at transmutation, and this in turn forced him into an antiprogressive stance. In the *Principles*, Lyell asked if the human species was to be "considered as one step in a progressive system by which as some suppose, the organic world advanced slowly from a more simple to a more perfect state."[77] No, said Lyell, because a complete discontinuity existed between humans and animals.

The superiority of man depends not on those faculties and attributes which he shares in common with the inferior animals, but on his reason, by which he is distinguished from them. . . . If this be admitted, it would by no means follow, even if there had been sufficient geological evidence in favour of the theory of progressive development that the creation of man was the last link in the same chain. For the sudden passage from an irrational to a rational animal is a phenomenon of a distinct kind from the passage from the more simple to the more perfect forms of animal organization and instinct. To pretend that such a step, or rather leap, can be a part of a regular series of changes in the animal world, is to strain analogy beyond all reasonable bounds.[78]

Lyell could make his case for the complete discontinuity between man and beast even stronger by arguing against progression in the animal kingdom as well. Lyell did not want natural history being used as evidence for the coming of humans because it would destroy the special status of man. Thus, his 1851 address was not just an argument against progression. Rather, its underlying purpose was to maintain the special status of humans, not by relying on some crude religious argument that resorted to miraculous creation, but by demonstrating that humans were not the "last term in a regular series of organic developments," because such a progressive series did not even exist.[79]

In Lyell's opening remarks, it is apparent that progressive development was inextricably linked to the coming of humanity. The recently published *Vestiges* had popularized the idea of transmutation, which Lyell had been arguing against for twenty years. He was not going to devote this address to refuting it again because Owen, Sedgwick, and Miller had done more than a sufficient job of discrediting the idea in their critiques of the *Vestiges*. While these men opposed transmutation, they had acceded to the idea of gradual development. They did not think that humans had evolved from a lower form, but they did believe that the fossil record indicated that the earth was gradually being prepared for the coming of humankind. Thus, rather than attacking transmutation, Lyell attacked progression. Before citing his evidence against progression, he quoted passages from Owen, Sedgwick, and Miller. There were many quotes Lyell could have chosen that would have illustrated the arguments made in favor of progression within the natural world; but each of the ones he chose specifically tied progression to the coming of humanity. From Sedgwick's fifth edition of *Discourse on the Studies of Cambridge*, he quoted the following:

> Are traces among the old deposits of the earth of an organic progression among the successive forms of life. . . . This historical development . . . of the form and functions of organic life during successive epochs seems to mark a gradual evolution of creative power, manifested by a gradual assent towards a higher type of being.[80]

According to Sedgwick, this progression to higher forms was not due to transmutation but to "creative additions," with man being the last. Owen made a similar claim. "Nature had advanced with slow and stately steps guided by the archetypal light amidst the wreck of worlds, from the first embodiment of the vertebrate idea, under its old icthyic vestment until it becomes arranged, in the glorious garb of the human form."[81] In *Footprints of Creation*, Hugh Miller claimed that the successive increase in the size of the brain in proportion to the spinal cord was evidence for Divine Plan. The brain of the fish was twice the size of the spinal cord (a ratio of 2:1), the reptile's two and a half times the size (2.5:1), the bird's triple (3:1), the mammal's four times the size (4:1), culminating in the appearance of "a brain that averages 23 to 1—reasoning, calculating man had come upon the scene."[82] Not only was this an argument for Divine Plan, but it used fossil evidence to argue for the coming of humanity. To discredit the argument, Lyell attempted to discredit the fossil evidence.

Although Lyell devoted most of the address to the animal and plant world, at the end he turned his attention to the appearance of humanity. In spite of arguing that other organisms had been present at all times in earth history, Lyell accepted the geological evidence for the recent appearance of humans on earth, because he did not want man included in the discussions about the history of organic life. But he claimed that even if man's origin had been much earlier by several periods, "the event would have constituted neither a greater nor less innovation on the previously established state of the animal world." For Lyell, the chasm between animals and humankind was absolute. No other animal even remotely approached humans in their intelligence or moral nature. Thus, the fossil record did not demonstrate a *scala naturae* leading to man. The mammals had not shown any significant advancement in all of the Tertiary that indicated a progression toward the human type. The appearance of humans would have to be treated as a complete discontinuity. By claiming that the recent origin of humans was a unique event, Lyell could then deny that the appearance of humans was evidence for a progressive system.

Lyell and Huxley continued to argue against progression in the 1850s, keeping track of new fossil evidence that could be used to support their views. For example, in 1855, Lyell wrote to Huxley regarding how Hugh Miller had shown how very reptilian the early ganoid fishes were.[83] But as the years went by, the position was becoming more and more untenable. Darwin's theory was a new assault on nonprogression and the special status of humanity. Lyell's species journals, particularly numbers three and four, document the struggle Lyell was having over progression, its relationship to transmutation, and the implications for the coming of humanity.

If progress be true we must look the whole prospect in the face. There will be found a gradation from the Quadrumanna to Man. Inferior Races of man will have proceeded, superior will follow. If we once imagine all the 4 types of Creation to stop because Man has appeared, we make his advent so great an event that the uniformity of the system is broken.[84]

But Lyell was still hoping he could deny progression. "The culminating point of Reptiles, is past, of Cycads & Gymnosperms [and] of fish," and "It seems now very generally admitted that the Reptile and Fish are not at the present epoch in the highest state of development, that they are, in this, passed [*sic*] their culminating point."[85]

Lyell wanted a naturalistic explanation for species coming and leaving, but as with Lamarck, he saw where Darwin's theory would lead. He wrote Huxley an anguished letter:

If we found all the leading classes, orders, families and genera or could reasonably hope to find them or could fairly infer that they did exist in the oldest periods, then we might by development get the species or I could conceive the genera in the course of millions of ages. But once admit the probable want of placental mammalia in the lower Silurian and we require such an event as the first appearance of that type at some subsequent period, an event which might compare with the first coming in of any other new type, ending with Man and it becomes difficult to know where to stop.[86]

He later admitted to Darwin that it was the implications for man that had set him against Lamarck. "As to Lamarck . . . I remember that it was the conclusion he came to about man that fortified me thirty years ago against the great impression which his arguments at first made on my mind."[87] How could Lyell accept progression and evolution, and still keep humanity's status special?

Both Lyell and Huxley wanted questions of theology kept distinct from those of natural history. But ironically, Lyell's deep-seated religious views regarding the dignity of man prevented him from ever fully embracing evolution, and this in turn forced him into his antiprogressionist stance. As Huxley wrote,

> I see no reason to doubt that, if Sir Charles could have avoided the inevitable corollary of the pithecoid origin of man—for which, to the end of his life, he entertained a profound antipathy—he would have advocated the efficiency of causes now in operation to bring about the condition of organic world, as stoutly as he championed that doctrine in reference to inorganic nature.[88]

Lyell reluctantly accepted Darwin's theory in 1863, but with a serious caveat—that of man. He still maintained that the chasm between human and animal was too great. Lyell later found a powerful ally in support of this view—Alfred Wallace. In 1869, in a review of the tenth edition of Lyell's *Principles*, Wallace argued that neither natural selection nor a general theory of evolution could "give an account whatever of the origin of sensation or conscious life . . . the moral and higher intellectual nature of man is as unique a phenomenon as was conscious life."[89] Such a view allowed Lyell to accept both progression and transmutation within the organic world, excluding humanity.

For Lyell, if humanity was to be kept special and distinct, then there couldn't be any progressive chain of organic development because the inevitable corollary was that humans were descended from brutes. However, the probable pithecoid ancestry of humans did not disturb Huxley. "It is not I who seek to base Man's dignity upon his great toe, or insinuate that we are lost if an Ape has a hippocampus minor."[90] Clearly, Huxley had different reasons than Lyell in continuing to advocate nonprogression.

Huxley and the Argument from Design

While Huxley had embraced the idea of transmutation immediately on the publication of the *Origin*, he did not accept progression until many years later. This was in spite of Lyell's
acknowledgment, in 1863, of progression (excluding humans)
in the organic world. Thus, although Lyell's influence on
Huxley was enormous, particularly in the 1850s, Lyell cannot
be the reason Huxley continued to deny progression in the
1860s. And while Huxley's belief in the type concept undoubtedly played a role in his rejection of progression, he still could
have argued that distinct persistent types progressively appeared
in the fossil record. Was Owen totally to blame for Huxley continuing to deny progression? I think not, for Owen's views were
more complex than Huxley suggests. Nevertheless, Huxley's
extreme animosity toward Owen gives us a clue to the underlying reason Huxley continued his antiprogressionist stance.

Leading scientists such as Buckland, Agassiz, Sedgwick, and
Owen believed that the progressive pattern of earth history was
evidence for the argument from design. However, even as early
as the 1840s, most geological debates were concerned with
interpreting particular fossil finds and trying to work out an
accurate stratigraphy.* They did not find the idea of secondary
causation incompatible with belief in a Creator. But Huxley's
obsession with keeping theological questions distinct from scientific ones led him to distort his opponents' views. For instance, nothing in Owen's critique of Lyell's address made even
the slightest reference to the argument from design. However,
Huxley's claim that Owen, Chambers, and others were letting
their theological beliefs influence their science served as a powerful rhetorical strategy for discrediting their views. Ironically,
as long as Huxley was unable to disassociate progression from
the argument from design, he continued to deny the increasing

---

*The classification, correlation, and interpretation of stratified rocks.

evidence in its favor. My evidence for this position is somewhat indirect; however, by accepting it as a working hypothesis, many of his views that seem contradictory become remarkably consistent.

Huxley's changing attitude toward transmutation becomes clearer if one examines it within the larger framework of keeping scientific theories free of theological implications. In Huxley's thinking, transmutation was very much tied to the idea of progression, which in turn was linked to the idea of Divine Plan. He had been totally against transmutation because of his belief in the type concept, which was the result of his own research and the work of Cuvier and von Baer. Yet Huxley enthusiastically accepted Darwin's theory in spite of having serious reservations about both natural selection and gradualism. But Darwin's theory was a totally *naturalistic* explanation of how species change, free from any supernatural influences. Although at the end of the *Origin*, Darwin acknowledges the Creator as being responsible for the laws of Nature, this does not change the fact that the overall argument in the *Origin* does not depend on the presence of a Creator. This one factor outweighed any difficulties that Huxley had with Darwin's mechanism of species change.

Huxley's enthusiasm for Darwin's theory contrasts sharply with his reaction to earlier theories of transmutation. Geoffroy St. Hilaire believed in the type concept and advocated a saltational view of species change. Superficially, Geoffroy's views appear to be much more similar to Huxley's than Darwin's. However, Geoffroy believed that the direct and sudden effect of the environment acting on fetal development could lead to changes in the adult that look like jumps. Fundamentally his mechanism was a teratological theory of evolutionary change.[91] Furthermore, his type concept was that of the transcendentalists, and Huxley had no use for their metaphysical speculations. While any mention of God or divine interference was absent from Lamarck's theory of transmutation, nevertheless,

Lamarck's scheme clearly reflected a belief in the Great Chain of Being, each organism evolving toward a higher and higher state of perfection. Huxley's linking of progression to transmutation and the idea of Divine Plan is most clearly seen in his 1854 review of the tenth edition of the *Vestiges*. The viciousness of this review was something even Huxley himself later regretted: "The only review I ever have qualms of conscience about, on the grounds of needless savagery, is the one I wrote on the *Vestiges*."[92] An analysis of the *Vestiges* review reveals that Huxley was not attacking the idea of transmutation; rather, he was objecting to a scientific theory being used to bolster the idea of Divine Creation.

Huxley was at his vitriolic best in his review of the *Vestiges*. Opening with a quote from *Macbeth*, "'Time was that when the brains were out, the man would die,'" Huxley did not understand why the *Vestiges* had not shared a similar fate. Describing the book as "a mass of pretentious nonsense" and a "notorious work of fiction," he was appalled that a tenth edition was being printed.[93] He devoted most of his attention to the fundamental proposition of the book, which was, quoting the author's own words: " 'The Natural proposition of the "Vestiges" is creation in the manner of law, that is the Creator working in a natural course or by natural means.' "[94] Huxley's analysis revolved around the definition of "law." He accepted the Vestigiarian's definition of law, which he also quoted:

> "Law . . . is merely a term of human convenience to express the orderly manner in which the will of God is worked out in external nature; and He must be ever present in the arrangements of the universe, as the only means by which they could be even for a moment sustained" (Proofs, lxii).[95]

Huxley paraphrased this definition as "Creation took place in an orderly manner, by the direct agency of the Deity." In doing so, he attempted to reduce Chamber's entire book to nothing more than another treatise on natural theology. For example, Huxley wrote, the Vestigiarian claimed that "natural laws" produced "winds" and "sometimes are concentrated . . . to produce hurricanes." Or " 'The Creator . . . is seen to have formed our earth, and effected upon it a long and complicated series of changes, in the same manner in which we find that he conducts the affairs of nature before our living eyes; that is, *in the manner of natural law.*' "[96]

Although the Vestigiarian claimed that his theory was different from those who used such vague phrases as "creative fiats," "interferences," or "interpositions of creative energy," for Huxley there was no difference. He asked

> If the author of the "Vestiges" really means by law, simply the mode in which the "Will of God"—who is ever present in the arrangements of the universe—takes effect . . . what meaning is there in passages we have just quoted? . . . If the Deity be ever present, and phenomena are the manifestation of his will—law being simply a name for the order in which these occur—what is every phenomenon, but the effect of a "creative fiat," an "interference," an "interposition of creative energy?"[97]

What appeared to be a scientific discussion on the organic world was nothing more than a elaborate version of the book of Genesis. Huxley dismissed the fundamental conception of the book as "totally inconsistent with itself—the product of coarse feeling operating in a crude intellect."[98]

According to Huxley, "creation by natural Law" was not a testable hypothesis, but he acknowledged that progression was a scientific proposition that could be accepted or rejected according to available evidence. He did not think the evidence supported progression. But even if "fully proved it would not be . . . an *explanation* of creation; such creation in the manner of natural law would . . . simply be an orderly miracle." Huxley believed "natural laws" were "nothing but an epitome of the observed history of the universe."[99] Thus, to claim progression was due to the Creator who worked "in the manner of natural law" was a meaningless statement. Huxley was trying to separate a scientific proposition (whether or not the fossil record was progressive) from metaphysical speculations (that the Creator worked in a lawlike fashion). But in the *Vestiges*, progression was inextricably linked with the presence of a Creator, and theological arguments were mixed with scientific ones. In addition, the book was filled with "blunders and mis-statements" giving a totally distorted view of the fossil record.[100] Thus, Huxley's wrath was directed not at the idea of transmutation, but rather at using scientific facts (which were often wrong) to promulgate a religious belief. Huxley's vehemence is doubly ironic. First, his antitheological rhetoric leaves him open to the same criticism he has of his opponents. And second, while most scientists claimed it was the poor science of the *Vestiges* that they objected to, underneath their professional criticism was the fear that the naturalistic explanation of "creation" threatened the special status of humanity.[101] Thus, while Huxley was claiming that the *Vestiges* was just another treatise in natural theology, merely an elaborate version of Genesis, most other scientists were worried that it would undermine their most deeply held religious views.

Huxley's argument changed somewhat in the 1860s. He argued not so much *against* progression as *for* persistent types. My claim is that in both cases the underlying reason remained the same. In Huxley's mind progressive development was associated with interpretations of the history of life that depended on the presence of a Creator. However, to continue to deny progression was becoming more and more difficult. It was true that as more and more fossils were found, the origin of each class was being pushed back to earlier and earlier times. The only problem was that the origin of *all* the classes was being pushed back, which meant that the appearance of the classes

remained successive. Invertebrates still appeared before vertebrates and fish still appeared before mammals, even though
they were all found much earlier in the fossil record than
anyone had previously imagined. While Darwin had shown that
a completely naturalistic account could explain the progressive
nature of the fossil record, nevertheless, progression had been
and still was being used as evidence for the argument from
design. Huxley needed a type of evidence that could be used
specifically in support of a naturalistic explanation of the history
of life. While Darwin was dismayed at Huxley's constant advocacy of persistence, in Huxley's mind persistent types were the
best sort of evidence in support of Darwin's theory because
other kinds of evidence could have other interpretations. Other
theories of species change as well as arguments against transmutation all claimed that progress was inevitable. Only
Darwin's theory allowed for both persistence and progression.
Huxley's antagonism toward natural theology was not a
byproduct of his acceptance of Darwin. Rather, his enthusiasm
for Darwin was in large part due to the absence of theology in
Darwin's theory.

In 1854, Huxley had cited approvingly Owen's work on
bone ossification in fish, but by 1862 the situation had
changed. Huxley's 1862 address was not just an attack on progression, but specifically assailed Owen's paleontological doctrines that claimed that the early ancestors in the fossil record
were more embryonic or more generalized. Darwin had said
much the same thing and wrote Huxley that the archetype was
in some degree embryonic and therefore capable of undergoing
further development.[102] For Owen these ancestors represented
the platonic archetype, and progressive development, which
eventually resulted in the appearance of humanity, was evidence
of the Creator. Owen's paleontology as espoused by Darwin
represented Huxley's dilemma in a nutshell. Huxley believed
that progressive development of the fossil record was powerful
support for the argument from design, particularly when advo-

cated by someone as influential as Owen, *even if* Owen did not explicitly refer to the design argument. While progressive development was certainly compatible with Darwin's theory, it could also be used in support of natural theology. Huxley had to find some type of evidence that helped distinguish Darwin's theory as a superior explanation of the fossil record from all others. That evidence eventually came in the form of Huxley's work on dinosaurs and horse phylogenies. But in 1862, Huxley believed that the best evidence was persistent types. If an insignificant amount of modification had taken place within any one group of plants or animals, this was

> quite incompatible with the hypothesis that all living forms are the result of a necessary process of progressive development. . . . Contrariwise, any admissible hypothesis of progressive modification must be compatible with persistence without progression, through indefinite periods.[103]

This suggests a rather more complex attitude on Huxley's part. He realized that Darwin's account provided an evolutionary explanation for how progressive development, stasis, and retrogression were all part of the history of life.

Huxley's research in the 1860s and 1870s on dinosaurs and fossil horses provided crucial support for Darwin's theory. As a result of that research, Huxley's views on progression and persistence changed once again. He finally acknowledged that many groups progressed because progression was no longer linked to the idea of Divine Plan. In addition, he no longer thought that persistent types were the best kind of evidence in support of Darwin's theory. This is most clearly seen in his 1876 American lectures on evolution.

Huxley was truly Darwin's bulldog on his trip to America, where he received a most enthusiastic welcome. According to one epistle contained in *The Life and Letters of Thomas Henry Huxley*, an American correspondent who had named his son

Thomas Huxley had commissioned Frederic Harrison to tell Huxley, "The whole nation is electrified by the announcement that Professor Huxley is to visit us next fall. We will make infinitely more of him than we did of the Prince of Wales and his retinue of lords and dukes." Huxley's first lecture (which was not specifically on evolution) was attended by two thousand people. His three lectures on evolution were printed in their entirety in the *New York Times* and later appeared in *Popular Science Monthly*.[104] The lectures were a brilliant and clear exposition of Darwin's theory in all its nuances and yet were easily understandable to the layperson. But as was true of many of Huxley's defenses of Darwin, he had a more fundamental message. Investigating the history of life should be regarded as a purely scientific question, free of theological speculations. Thus, he told his audience,

> [M]y present business in not with the question as to how Nature has originated, as to the causes which have led to her origination, but as to the manner and order of the appearance of natural objects. . . . This is a strictly historical question, a question as completely historical as that about the date at which the Angles and the Jutes invaded England. But the other question about creation is a philosophical question, and one which cannot be solved or even approached by the historical method.[105]

Huxley contrasted Darwin's theory with other theories about the history of life. In a clever rhetorical move, he did not try to discredit the biblical hypothesis of special creation. Rather, he attacked what he referred to as the Miltonic hypothesis, because this was a hypothesis that could be tested empirically. Did the order of appearance of organisms in *Paradise Lost* correlate with the order found in the fossil record? In his typically acrid style, Huxley conclusively demonstrated that the fossil evidence contradicted the order that animals appeared in the Miltonic hypothesis. Allowing for a loose interpretation of what "day" meant in the Miltonic account, the order of appearance in the fossil record should parallel what was described in *Paradise Lost*. Thus, birds should be found in the Devonian, Silurian, and Carboniferous. But birds did not appear until far more recently. Huxley pointed out several other inconsistencies, the most dramatic being the fish. According to Milton, the fish and whales appeared on the fifth day, and thus we should be able to find remains of fish in the rocks preceding the Carboniferous. While plenty of fish existed, they were different from the ones existing today, and no whales were present. Since no modern-day fish remains had been found in the older rocks,

> you are introduced again to the dilemma that either the crea-
> tures which were created then, which came into existence the
> sixth day, were not those which are found at present, are not
> the direct and immediate predecessors of those which now
> exist; in which case you must either have had a fresh creation
> of which nothing is said, or a process of evolution; or else the
> whole story must be given up, as not only devoid of any cir-
> cumstantial evidence, but contrary to that evidence.[106]

To salvage the Miltonic hypothesis, the only alternative was the idea of miraculous creation. But ultimately such arguments relied on faith. And as Huxley emphasized throughout the lectures, there could never be scientific proof for special creation.

Having effectively demolished the Miltonic hypothesis, he then moved on to the various kinds of evidence in support of Darwin's theory. He reminded his listeners that "it is quite hopeless to look for testimonial evidence of evolution" because the nature of the case precludes such evidence. "Our sole inquiry is, what foundation circumstantial evidence lends to that hypothesis, or whether it lends any, or whether it contro-verts it; and I should deal with the matter entirely as a question of history."[107] With what was an obvious attack on the natural theologians, he claimed that "I shall not indulge in the discussion of any speculative probabilities. I shall not attempt to show that Nature is unintelligible unless we adopt some such hypothesis."[108] Huxley was going to defend the evolutionary hypothesis, purely on the evidence, just as he had discredited the Miltonic hypothesis by showing that the fossil record contradicted such an account.

Huxley divided the evidence into three types. The first kind was neutral, neither helping nor inconsistent with the evolution hypothesis. The second type indicated a strong probability; the third type, he believed, was "as complete as any evidence which we can hope to obtain upon such a subject, and being wholly and entirely in favor of evolution, may be fairly called demonstrative evidence of its having occurred."[109] In this scheme, persistent types had become far less important. They were in the category of neutral evidence. The relationship of birds to dinosaurs belonged to the second category, while the most compelling evidence for Huxley was the many fossil horse remains that Othniel C. Marsh had found, which will be discussed in detail shortly.

Huxley described many cases of persistence, such as the marine invertebrate lamp-shells. Two species, *Terebratula* and *Globerina*, showed no significant modification since the chalk period of the Cretaceous. The *Beryx* was a highly differentiated fossil fish from the chalk that looked very similar to some present-day fish found in both the Atlantic and the Pacific. And going even further back into the Carboniferous, remains of scorpions had been found that were almost indistinguishable from present-day ones. Mollusks belonging to the group *Lingula* that were presently found along the shores of Australia were similar to ones found in the Cambrian.[110] Huxley also mentioned the Mesozoic reptiles *Ichthyosauria* and *Plesiosauria*, which eventually disappeared, but "throughout the whole of that great series of rocks they present no important modifications."[111] Finally, Huxley mentioned the Permian lizards, which "differ astonishingly little from the lizards which exist at the present day."[112] He admitted that persistent types provided evidence that was indifferent; in other words, "they may afford no direct support to the doctrine of evolution."[113] He knew that many had interpreted his advocacy of persistence as an argument against evolution. But he explained to his audience why this was not the case. Darwin's theory was based on two great factors: the tendency to vary and the influence of the environment on "the parent form and the variations which are thus evolved from it."[114] While the cause of the variations were not known, nevertheless, simple observation confirmed that the tendency to vary was found throughout the organic world. If the environmental conditions were such that the parental form was better suited than some of the derived forms, it would maintain itself and the derived forms would be exterminated. However, if the conditions were better suited for the derived form, the parent form would die out and the derived form would take its place. "In the first case, there will be no progression, no advance of type, through any imaginable series of ages; in the second place there will be modification and change of form. Thus, the existence of these "persistent types of life is no obstacle in the way of the theory of evolution."[115]

Persistent types did not provide what he considered "demonstrable proof" of evolution. Rather, they were in the category of neutral evidence, providing neither direct proof nor an obstacle to Darwin's theory. However, persistent types still played an essential role in Huxley's ongoing battle with those who insisted on mixing theology with science. By 1876, evolution in the general sense of the word was reasonably well accepted. However, what was particularly popular was the idea of designed evolution.[116] Mivart, Owen, and the Duke of Argyll all reformulated the idealist version of the argument from design to show that Darwin's purely utilitarian concept of adaptation was not a complete explanation.[117] But this was precisely the sort of argument that Huxley was trying to discredit. Persistent types undermined the argument from design because they showed there was no intrinsic drive toward perfection. As Huxley claimed, they were "fatal to any form of the doctrine of evolution, which necessitates the supposition that there is an intrinsic necessity on the parts of animal forms which once come into existence to undergo modification."[118] By the 1870s, Huxley accepted that many groups showed progression, but he never abandoned his belief in persistent types. Thus, a belief in persistent types was not synonymous with a denial of progression. Some types persisted while others progressively changed. In modern parlance, Huxley recognized stasis in the fossil record, and he deserves credit for drawing attention to this fact. Indeed, Huxley's views on progression sound remarkably modern. But were they?

Huxley and the Modern View of Progression

The problem of progression in the fossil record is a tricky one. The fossil evidence for progression was far from clear cut in the 1830s, 1840s, and even through part of the 1850s. Today, no one denies that particular groups of organisms have appeared at different times and individual species have become extinct and new ones have come into being. However, virtually all major groups were present by the Cambrian. And does the record show progression, if we define progression as an increase in complexity? Is the modern horse more complex than *Eohippus?* Are dinosaurs simpler in organization than modern lizards? If one takes any chunk of the fossil record, it is quite difficult to demonstrate progression within it.[119] This was even more difficult to do in Huxley's time when sampling of the fossil record was far less systematic than today. In spite of accepting evolution, Huxley was suspicious of the progressionist interpretation embodied in Darwin's scheme, maintaining that the record did not demonstrate a connection between evolution and increasing levels of organization. This sounds very similar to the position espoused by many present-day paleontologists. Was Huxley ahead of his time in thinking about progression? Peter Bowler presents Huxley's antiprogressionism very favorably, claiming that belief in progression was an outdated notion from the early part of the nineteenth century. "Huxley points the way toward the increasing suspicion of biological progressionism that has grown up in the twentieth century as the earlier period's optimistic faith in general progress has crumbled."[120] Bowler even asserts that Huxley's views "mark the beginning of the modern attitude whereby evolutionists simply ignore the whole issue of progression."[121] As much as I would like to agree with Bowler's analysis, I think it is incomplete.

Bowler's favorable account of Huxley is based on Huxley's work on dinosaurs and fossil horses. He claims that Huxley's work showed that the "old linear concept of development was unrealistic."[122] Paraphrasing Gould, the pattern of evolution is a bush, not a ladder. However, I think the primary significance of the dinosaur work was that it was the key factor in Huxley giving up his belief in saltation. Huxley's work on horses continued to provide evidence for the general theory of evolution and specifically for gradualism.[123] The tracing of phylogenies of the dinosaurs and horses also showed the multilinearity of evolutionary development. There was no single line of development. The dinosaur work was particularly important in that it showed that two "higher" groups, birds and mammals, arose separately from two quite distinct lines of reptiles. Thus, the early progressionists' idea that birds were a step on the way toward mammals was clearly incorrect. As more and more fossils turned up, as the evidence for Darwin's theory continued to mount, Huxley finally acknowledged that the successive appearance of various classes did not mean such a progressive scheme would always be linked to the idea of Divine Plan.

Certainly Huxley had no use for the Chain of Being, an outdated notion that was prevalent in the eighteenth and early nineteenth centuries. But many advocates of discontinuous progression believed in the sequential appearance of the various classes while admitting that it was impossible to demonstrate an increase in level of organization within classes. They also realized that it was often difficult to draw comparisons between classes. In this respect, one could claim that the paleontology of Agassiz and Owen was modern as well. However, they also believed that the progressive nature of the history of life demonstrated the Creator's Divine Plan. This was the fundamental difference between Huxley and most nineteenth-century advocates of progressionism. Huxley's views were modern in that he realized that natural theology had no place in modern scientific thought. But having such a strong opinion on the matter caused him to deny the validity of much excellent work done by paleontologists in the service of natural theology. As Bowler pointed out, "Their efforts to work out a system that was both scientifically and theologically acceptable led to a notable expansion of the fossil record which had enormous impact on our understanding of the history of life."[124] Perhaps Darwin's real genius lay in his ability to look at the works of natural theology and distill from them those ideas and "facts" to construct an argument that was totally free from design. In Darwin's theory, natural selection plays the same role as the Creator. Darwin believed there was "no limit to the amount of change, to the beauty and infinite complexity of the coadaptations between all organic beings, one with another and with their physical conditions of life, which may be effected in the long course of time by nature's power of selection."[125] For Huxley, finally, here was a theory of transmutation that provided a totally naturalistic explanation for the pattern and processes of life.

In struggling to free science from theological implications, Huxley let his own philosophical beliefs influence his interpretation of the data. However, Huxley was certainly not unique in this respect. Like the creationists he despised, he played a crucial role in the debates over progression in the fossil record and its relationship to evolutionary theory. Why he maintained such a strong antiprogressionist stance for such a long time was due to a variety of factors. Undoubtedly, just the sheer inertia of ideas played a role. He was committed to a theory of type and was heavily influenced by von Baer, who argued that one could not rate the different types as being higher or lower than the others. By the mid-1850s his animosity toward Owen had become extreme and he tried to discredit the man whenever possible. Yet, as I pointed out, he also was more than willing to cite Owen's early work when it suited his needs. Therefore, I believe the crucial factor in Huxley's eventual acceptance of progression was that he finally disassociated it from the idea of Divine Plan. This happened gradually through the 1860s and 1870s as more and more fossil finds provided support for Darwin's theory. In evaluating this new evidence that supported gradualism, Huxley also realized that progression was an intrinsic part of Darwin's theory.

> The hypothesis of evolution supposes that at any given period in the past we should meet with a state of things more or less similar to the present, but less similar in proportion as we go back in time. . . . [I]f we traced back the animal world and the vegetable world we should find preceding what now exist animals and plants not identical with them, but like them, only increasing their differences as we go back in time, and at the same time becoming simpler and simpler until finally we should arrive at that gelatinous mass which, so far as our present knowledge goes, is the common foundation of all life.[126]

In concluding his first lecture to the Americans, Huxley told them, "The hypothesis of evolution supposes that in *all this vast progression* [emphasis added] there would be no breach of continuity, no point at which we could say 'This is a natural process,' and 'This is not a natural process.' "[127] Huxley now argued that the new fossil finds that showed a gradual progressive scheme provided the most convincing evidence for the truth of the Darwinian hypothesis. Therefore, progression was no longer linked to Divine Plan. The next chapter examines the work that led to him accepting progression and, more importantly, to his abandonment of his belief in saltation.

Notes

1. Jacob Gruber, January 10, 1989, personal communication.
2. James Hutton, *Theory of the Earth, Transactions Royal Society of Edinburgh* 1 (1788), p. 304.
3. Martin Rudwick, "Uniformity and Progression: Reflections in the Structure of Geological Theory in the Age of Lyell," in *Perspectives in History of Science and Technology*, Duane Roller, ed. (Norman: University of Oklahoma Press, 1971), p. 214.
4. Ibid., pp. 209–27.
5. See Martin Rudwick, *The Great Devonian Controversy* (Chicago: University of Chicago Press, 1985), pp. 44–45, for a discussion of the relationship of the scriptural geologists to the scientific geologists.
6. Peter Bowler, *Fossils and Progress: Paleontology and the Idea of Progressive Development in the Nineteenth Century* (New York: Science History Publications, 1976), p. 76.
7. I realize that there is some controversy over this position today. Stephen Gould in particular argues that Darwin did not believe in progress. But I believe Gould's position is a minority one and he should not project backwards to the nineteenth century the sophisticated debates that are going on in the evolutionary biology community today regarding notions of progression and complexity. While Darwin realized that his theory did not have to lead to progress, he nevertheless accepted that overall the fossil record showed that the pattern of life was one of divergence with increasing complexity. See *Evolutionary Progress*, Matthew Nitecki, ed. (Chicago: University of Chicago Press, 1988), for a good summary of the current debates over the notion of progress and evolution in both the nineteenth and twentieth centuries. See also Bowler, *Fossils and Progress*, chapter 6; Robert Richards, *The Meaning of Evolution* (Chicago: University of Chicago Press, 1992); and Michael Ruse, *From Monad to Man* (Cambridge: Harvard University Press, 1997).
8. Charles Darwin to Reverend Rodwell. Rodwell told Francis Darwin that he remembered Charles saying this to him. *Life and Letters of Charles Darwin* (*LLCD*), 2 vols., Francis Darwin, ed. (London: John Murray, 1887), 2: 140.

9. Charles Lyell, *The Principles of Geology*, 3 vols. (London: John Murray; Chicago: University of Chicago Press, 1991), 1: 165–85.

10. Charles Lyell, "Anniversary Address of the President," *Quarterly Journal of the Geological Society* 7 (1851): xxv–lxxvi.

11. Ibid., p. liv.

12. Ibid., p. xxxvi.

13. Ibid.

14. Ibid., pp. xxxv–xxxvii.

15. Ibid.

16. Ibid., pp. xxxvii–xxviii.

17. Ibid., pp. lxxv–lxxvi.

18. Huxley would later date this as belonging to the New Red Sandstone, but by that time both Lyell and Huxley had given up their antiprogressionism for different reasons. See Bowler, *Fossils and Progress*, p. 75.

19. Lyell, "Anniversary Address," p. lx.

20. Ibid., pp. lxvi–lxvii.

21. Ibid., p. xxxviii.

22. Gideon Mantell, *The Wonders of Geology*, quoted in Bowler, *Fossils and Progress*, p. 77.

23. Edward Forbes, "On the Manifestation of Polarity in the Distribution of Organic Beings in Time," *Notices of Proceedings of Royal Institute* (1851–1854): 429, quoted in Bowler, *Fossils and Progress*, p. 77.

24. Bowler, *Fossils and Progress*, p. 78. See Constance Prévost, *Comptes rendus* 20 (1845): 1062–71.

25. Ibid.

26. Lyell, "Anniversary Address," p. xxxvi. See Constance Prévost, *Comptes rendus* 31 (1850): 461, and A. D. D'Orbigny, *Comptes rendus* 30 (June 1850): 807.

27. Thomas Huxley, "Vestiges of the Natural History of Creation Tenth Edition London, 1853," *The British and Foreign Medico-Chirurgical Review* 13 (1854): 425–39, *Scientific Memoirs of Thomas Henry Huxley* (*SMTHH*), 4 vols., Michael Foster and E. Ray Lancaster, eds. (London: Macmillan & Co., 1898–1902), suppl., pp. 1–19.

28. Ibid., p. 7.

29. Ibid., p. 8.

30. Ibid., p. 10.

31. Richard Owen, *The Hunterian Lectures in Comparative Anatomy*, Philip Sloan, ed. (Chicago: University of Chicago Press, 1992).

32. Huxley, "Vestiges," *SMTHH*, p. 11.

33. Ibid., Huxley quoting Owen, p. 11.

34. Anonymous, "Lyell—On Life and Its Successive Development," *Quarterly Review* 89 (1851): 412–51.

35. Thomas Huxley, November 9, 1951, *Life and Letters of T. H. Huxley* (*LLTHH*), 2 vols., Leonard Huxley, ed. (New York: D. Appleton & Co., 1872), 1: 101.

36. "Lyell—On Life," p. 427.

37. Ibid., p. 450.

38. Ibid., p. 430.

39. Thomas Huxley, "On the *Stagonolepis Robertsoni* (Agassiz) of the Elgin Sandstones," *Quarterly Journal of the Geological Society* 15 (1859): 440–60, *SMTHH* 2: 117–19.

40. Herbert Spencer, December 16, 1858, *Life and Letters of Herbert Spencer*, 2 vols., Duane Ducan, ed. (London: Methuen and Co., 1908), 1: 91.

41. Thomas Huxley, 1887, "On the Reception of *The Origin of Species*," *LLCD* 1: 538.

42. Ibid., p. 542.

43. Thomas Huxley, "On the Persistent Types of Animal Life," *Proceedings of the Royal Institution of Great Britain* 3 (1859): 151–53, *SMTHH* 2: 91.

44. Ibid.

45. Charles Darwin to Asa Gray, May 22, 1860, *LLCD* 2: 104.

46. See Darwin, "Notebook E," MS p. 95. In *Charles Darwin's Notebooks, 1836–44,* Paul Barrett et al., eds. (Ithaca, N.Y.: Cornell University Press, 1987), pp. 422–23. See also Richards, *Meaning of Evolution*, pp. 86–87.

47. Thomas Huxley, 1862, "Geological Contemporaneity and Persistent Types of Life," in *Collected Essays of Thomas Huxley*, vol. 8, *Discourses Biological and Geological* (London: Macmillan & Co., 1908), p. 303.

48. Charles Lyell to Leonard Horner, February 23, 1862, *Sir Charles Lyell: Life, Letters and Journals*, 2 vols., Katherine Lyell, ed. (London: John Murray, 1881), 2: 356.

49. Thomas Huxley, May 6, 1862, *LLTHH* 1: 221.

50. Charles Darwin, May 10, 1862, *More Letters of Charles Darwin* (*MLCD*), 2 vols., Francis Darwin and A. C. Seward, eds. (London: John Murray, 1903), 2: 234.

51. See Adrian Desmond, *Archetypes and Ancestors* (Chicago: University of Chicago Press, 1984), p. 93 for a further discussion of this point.

52. I will discuss Huxley's conversion to gradualism in chapter 5.

53. Thomas Huxley, 1870, "Paleontology and the Doctrine of Evolution," in *Discourses Biological and Geological*, p. 347.

54. See Thomas Huxley, "On New Labyrinthodonts from the Edin-

burgh Coal-Field," 1862, *SMTHH* 2: 530–35; "Description of *Anthracosaursu Russelli*, a New Labyrinthdont from the Lanarkshire Coal-Field," 1863, *SMTHH* 2: 558–72; "On a New Species of *Telerpeton Elginense*," 1867, *SMTHH* 3: 205–13; "On *Saurosternon Bainii*, and *Pristerodon McKay*, Two new Fossil Lacertilian Reptiles from South Africa," 1868, *SMTHH* 3: 298–302.

55. Huxley, "Paleontology and the Doctrine of Evolution," p. 348.

56. Ibid., p. 349.

57. Bowler, *Fossils and Progress*, p. 75, and Nicholaas Rupke, *Richard Owen, Victorian Naturalist* (New Haven: Yale University Press, 1994), pp. 149–50. Owen's reinterpretation followed Logan's second paper in 1852 on the footprints.

58. Thomas Huxley, "On Hyperodapedon," *Quarterly Journal of the Geological Society* 25 (1869): 138–52, *SMTHH* 3: 390.

59. Peter Bowler correctly distinguishes two concepts of design. The first was William Paley's idea of designed adaptation, while the second grew out of the idealist movement. Darwin's theory may have been explicitly a reaction against the idea of designed adaptation, but it equally undermined the idea of the idealist version of design. Huxley had no use for either concept because both ultimately relied on the presence of a Creator in their explanation of the natural world. See Peter Bowler, "Darwinism and the Argument from Design: Suggestions for a Reevaluation," *Journal of the History of Biology* 10 (1977): 29–33.

60. See Peter Bowler, *Evolution* (Berkeley: University of California Press, 1984), pp. 55–59, for a discussion of their work.

61. Ibid., p. 60.

62. William Buckland, *Reliquiae Diluvianae* (London: John Murray, 1823), p. iii.

63. Ibid., p. 224.

64. William Buckland, *Geology and Mineralogy Considered with Reference to Natural Theology* (Philadelphia: Phillip Carey, Lea and Blanchard, 1837), p. 436.

65. Ibid., p. 66.

66. Ibid., p. 90.

67. Ibid., p. 436; Buckland was quoting *British Critic* 17 (January 1834): 194.

68. Buckland, *Geology and Mineralogy*, p. 441.

69. Bowler, *Fossils and Progress*, pp. 28–39.

70. Ibid., p. 39.

71. Lyell, 1832, *Principles of Geology*, 2: 325–26.

72. Michael Bartholomew, "Lyell and Evolution: An Account of Lyell's Response to the Prospect of an Evolutionary Ancestory for Man," *British Journal for the History of Science* 6 (1973): 261–303.

73. Ibid., p. 264.

74. Charles Lyell to Gideon Mantell, March 2, 1827, *LLCL* 1: 168.

75. Ibid.

76. Bartholomew, "Lyell and Evolution," p. 264.

77. Lyell, *Principles*, 1: 155.

78. Ibid.

79. Lyell, "Anniversary Address," p. xxxix.

80. Quoted in ibid., p. xxxiii.

81. Quoted in ibid., p. xxxiv.

82. Quoted in ibid., p. xxxv.

83. Charles Lyell, August 29, 1855, Huxley Manuscripts (HM) 6: 9

84. Charles Lyell, August, 2, 1858, Leonard Wilson, ed., *Sir Charles Lyell's Scientific Journals on the Species Question* (New Haven, Conn.: Yale University Press, 1970), pp. 190–91.

85. Ibid., p. 192.

86. Lyell to Huxley, June 17, 1859, HM 6: 120.

87. Lyell to Darwin, March 15, 1863, *LLCL* 2: 365.

88. Huxley, 1887, "On the Reception of *The Origin of Species*," *LLCD* 1: 546–47. In the footnotes Huxley quoted letters of Lyell to various people that back up Huxley's claim.

89. Alfred Russel Wallace, "Geological Climates and the Origin of Species," *Quarterly Review* 126 (1869): 391.

90. Thomas Huxley, 1863, *Collected Essays of Thomas Huxley*, vol. 7, *Man's Place in Nature* (New York: D. Appleton & Co., 1898), p. 152. While this statement is clearly an attack on Richard Owen (see chapter 6), it nevertheless demonstrates that an evolutionary account of human origins was not problematic for Huxley.

91. Toby Appel, *The Cuvier-Geoffroy Debate* (New York: Oxford University Press, 1987), pp. 131–34.

92. Huxley, "On the Reception of the *Origin of Species*," *LLCD* 1: 542.

93. Huxley, "Vestiges," *SMTHH*, suppl., p. 1.

94. Ibid. (Huxley quoting the "Vestiges").

95. Ibid., p. 3

96. Ibid., p. 4 (Huxley quoting p. 113 of the *Vestiges*).

97. Ibid., p. 5.

98. Ibid.

99. Ibid.

100. Ibid., p. 2.

101. See Martin Rudwick, *The Meaning of Fossils* (Chicago: University of Chicago Press, 1985), pp. 205–207, for a discussion of the reception of the *Vestiges*.

102. Charles Darwin, April 23, 1854, *MLCD* 1: 73.

103. Thomas Huxley, "Geological Contemporaneity," in *Discourses Biological*, p. 306.

104. Lizzie Huxley, *LLTHH* 1: 492, 501.

105. Thomas Huxley, "The Three Hypotheses of the History of Nature," *Popular Science Monthly* 10 (1877): 51.

106. Ibid., p. 54.

107. Ibid., p. 55.

108. Ibid.

109. Ibid., p. 56.

110. Thomas Huxley, "The Negative and Positive Evidence," *Popular Science Monthly* 10 (1877): 208–209.

111. Ibid., p. 209.

112. Ibid., p. 211.

113. Ibid., pp. 210–11.

114. Ibid., p. 210.

115. Ibid.

116. See Alvar Ellegard, *Darwin and the General Reader: The Reception of Darwin's Theory of Evolution in the British Periodical Press, 1859–1872* (Goteburg, Germany: Acta Universitatis Gothenburgensis, 1958).

117. See Bowler, "Darwinism and the Argument from Design," pp. 36–42, for a more complete discussion of how the argument from design was reformulated after the publication of the *Origin*.

118. Huxley, "The Negative and Positive Evidence," pp. 209–10.

119. David Raup, March 1987, personal communication. Stephen Gould is also a strong proponent of the idea that the fossil record does not show any significant increase in complexity after the Cambrian age.

120. Bowler, *Fossils and Progress*, p. 132. Since Bowler in a more recent book, *The Non-Darwinian Revolution*, classifies Huxley as a pseudo-Darwinian, it is unclear whether he still has such a laudatory opinion of Huxley in relation to the question of progression and the fossil record.

121. Bowler, *Fossils and Progress*, p. 139. Whether modern evolutionists ignored the question of progression is certainly debatable. Julian Huxley was a strong believer in evolutionary progress, and he was able to find ample

support for his views from the paleontological literature of the twentieth century. See Marc Swetlitz, "Julian Huxley and the End of Evolution," *Journal of the History of Biology* 28 (1995): 181–217. Furthermore, interest in progress and evolution remains high. See Nitecki, *Evolutionary Progress*, as well as Ruse, *From Monad to Man*.

122. Bowler, *Fossils and Progress*, p. 132.

123. I will be discussing Huxley's research on dinosaurs and horses in greater detail in chapter 5.

124. Bowler, *Fossils and Progress*, p. 28.

125. Charles Darwin, *The Origin of Species* (1859; New York: Avenel, 1976), p. 153.

126. Huxley, "The Three Hypotheses," p. 47.

127. Ibid.

Chapter Five

From Saltation
to Gradualism

*Darwin's position might . . . have been even stronger than it is
if he had not embarrassed himself with the aphorism "Natura
non facit saltum" which turns up so often in his pages. We
believe . . . that Nature does make jumps now and then and a
recognition of that fact is of no small import in disposing of
many minor objections to the doctrine of transmutation.*

Thomas Huxley, 1860[1]

*There is conclusive evidence that the present species of ani-
mals and plants have arisen by gradual modification of
preexisting species.*

Thomas Huxley, 1874[2]

I have argued in previous chapters that Huxley's belief in
saltation made sense in light of his early views on archetypes.
Cuvier's research and the work of von Baer, as well as Huxley's
own research, demonstrated that organisms could be grouped
into distinct types, each following its own unique pattern of
development. The fossil record did not document a graduated
chain connecting one group of organisms to another. Rather,
groups became extinct and new forms abruptly appeared with
no evidence of transitional forms. Finally, Lyell, while a gradualist in geology, was a saltationalist in his thinking about the bio-

logical world. Saltation allowed Huxley to accept evolution while maintaining his belief in the concept of type. He believed that *natura facit saltum* was simply a more accurate description of the history of life. Indeed, he cautioned Darwin on the eve of the publication of the *Origin*, "You have loaded yourself with an unnecessary difficulty in adopting *natura non facit saltum* so unreservedly."[3] Darwin was well aware of the problems that the fossil record presented for his gradualist scheme. He did not deny the difficulties, but argued that the gaps in the fossil record were due to the irregular process of fossilization, competitive exclusion, and migrations into new areas. Whether an organism was actually preserved or not depended on so many different factors that the fossil record represented a quite incomplete chronicle of the history of life. Moreover, this was the reason that it was so rare to see transitional forms between many present-day groups. As he wrote in the *Origin*,

> We may thus account even for the distinctness of whole classes from each other—for instance of birds from all other vertebrate animals—by the belief that many animal forms of life have been utterly lost through which the early progenitors of birds were formally connected with early progenitors of the other vertebrate classes.[4]

Of the many examples Darwin could have picked to illustrate the problem of the lack of transitional forms between present-day species, it is ironic that he chose the one with which Huxley was to become intimately involved and one which was key to Huxley abandoning his saltationalist views: the relationship of birds to reptiles. Huxley's work on *Archaeopteryx* and dinosaurs led him to propose dinosaurs as the antecedent form between reptiles and birds.

Huxley's dinosaur work was in one sense a continuation of research he had done on reptilian fossils that had begun prior to the publication of the *Origin*. He published nine monographs on fossils between 1857 and 1859. Although much of his work involved fossils, he claimed they did not really interest him. In 1854, Huxley was offered Edward Forbes's post of paleontologist and lecturer in natural history at the Geological Survey of Mines. He flatly refused the paleontologist position and accepted the post of lecturer only provisionally. He wrote in his autobiography that he told Sir Henry Foster, "I did not care for fossils and that I should give up the natural history as soon as I could get a physiological post."[5] What began as a search for evidence that supported nonprogression and persistent types resulted eventually in work that Huxley considered undeniable proof of evolution. As he was later to write, "If the doctrine of Evolution had not existed paleontologists must have invented it, so irresistibly is it forced upon the mind by the study of the remains of the Tertiary mammalia which have been brought to light since 1859."[6]

Mario di Gregorio claims that Ernst Haeckel, rather than Darwin, was primarily responsible for Huxley's incorporation of evolutionary ideas in his own research. Di Gregorio cites Huxley's 1868 memoir "On the Animals Which Are Most Nearly Intermediate between Birds and Reptiles" as the first time Huxley explicitly used the notion of evolution in a monograph.[7] But such a claim does not hold up to analysis. The previous chapter demonstrated that Huxley argued for the prevalence of persistent types specifically because in 1859 and the early 1860s he believed persistent types were the best evidence to distinguish the Darwinian hypothesis from other evolutionary schemes that involved the argument from design. His correspondence with Lyell also indicates that Huxley was already thinking along evolutionary lines. In addition, Huxley's most famous work, *Man's Place in Nature* (1863), was inspired by the *Origin*.[8] Clearly, Huxley had been thinking about the implications of evolutionary theory before Haeckel's descent theory.

Certainly Haeckel played an important role in Huxley's renewed interest in fossils. Haeckel published his *Generelle Morphologie der Organismen* in 1866, and Huxley met with him that same year in London. Haeckel's claim that Huxley was "deeply interested" in his daring endeavor is supported by Huxley's letter to Haeckel:

> The main thing about which I am engaged is a revision of the Dinosauria—with an eye to the Descendez Theorie! The road from Reptiles to Birds is by way of Dinosauria to the Ratitae —the Bird "phylum" was Struthious, and wings grew out of rudimentary fore-limbs. You see that among other things I have been reading Ernst Haeckel's *Morphologie*.[9]

But descent theory was not the only idea of Haeckel's that Huxley found attractive. Huxley, several years earlier, had written Lyell that he thought transmutation of species might occur in a way analogous to chemical transformation, with one atom substituting for another leading to very different organic compounds (the letter has been quoted in chapter 3). Huxley found validation for this idea in Haeckel's *Generelle Morphologie*.

> Phylogeny, or the paleontological development of blood-related forms which lead to the establishment of genuses, orders, families, and all the other categories of the organic world, is a physiological process which, like all other physiological functions of organisms, proceeds of absolute necessity by mechanistic means.
>
> These means are the atomic and molecular motions which piece organic matter together . . . first in the concatenation of organic matter and then in the subtle compounds of which the active plasma of the constituent plastides of all organisms is composed. The phylogenetic or paleontological development . . . is neither the premeditated purposeful result of a creative intelligence nor yet the outcome of some mysterious vital force of nature, but rather the simple and necessary operation of that familiar physical and chemical process identified by physiology as the operative mechanism in the development of organic matter.[10]

Haeckel's theory of descent was especially attractive to Huxley on two counts. First, it was based on interpreting von Baer's embryological claims in terms of evolutionary theory, and second, it was mechanistic. In the above quote, Haeckel specifically pointed out that phylogenetic development was "neither the premeditated purposeful result of a creative intelligence nor the outcome of some mysterious vital force"—resonating with Huxley's own philosophy. Darwin's own version of descent had these features as well. Huxley was as open to Haeckel as he was for precisely the same reasons he found Darwin attractive and specifically because Darwin and von Baer had already significantly influenced his thinking.

While Huxley actively promulgated Darwin's theory in the early 1860s, he nevertheless maintained that Darwin's insistence on gradualism was not supported by the evidence. As I pointed out in chapter 3, Huxley was not particularly interested in finding transitional forms because they would undermine his deeply held belief in the concept of type. Fossils were being found at a rapid rate, pushing the origin of classes back to earlier and earlier epochs. Many of these new fossils supported Huxley's idea of persistent types, but some also appeared to be transitional forms. Since the fossil record had been the primary reason for Huxley's denial of gradualism, it is not surprising that new fossil finds would be crucial to his abandoning saltation.

Birds, Reptiles, and Dinosaurs

In 1861, a remarkable fossil was found in the Jurassic limestone at Solnhofen in Bavaria—what appeared to be a feathered bird. Along with other fossils from the Solnhofen, it was purchased in 1862 by the British Museum. Richard Owen, who was head of the natural history collections, lost no time in preparing a description of it, which he read before the Royal Society in November. Naming it *Archaeopteryx macrurus*, Owen claimed that "the least equivocal parts of the present fossil declare it to be a Bird, with rare peculiarities indicative of a distinct order in that class."[11] The Mesozoic bird had retained certain primitive structures that were only "embryonical and transitory in modern representatives of the class and consequently [showed] a closer adhesion to the general vertebrate type."[12] As early as 1863, Huxley was teaching in his classes that birds were an extremely modified and aberrant reptilian type. However, he did not immediately jump on the evolutionary implications of *Archaeopteryx* as a transitional organism. Adrian Desmond suggests this was because he was still caught up in Lyellian non-progression, believing that birds were flourishing in Paleozoic times. If that were true, Jurassic *Archaeopteryx* was far too recent to be considered one of the earliest members of its class.[13] Instead, it might be an example of a persistent type that had eventually become extinct. But we have no evidence that Huxley ever considered *Archaeopteryx* an example of a persistent type. As I have said, finding transitional organisms would not have been a priority for Huxley in the early 1860s because of his belief in both the type concept and nonprogression. More importantly, at the time of *Archaeopteryx*'s discovery, Huxley was heavily involved with the relationship of apes to humans and developing the lectures that eventually appeared as *Man's Place in Nature*. The most controversial aspect of Darwin's theory was its implications for human origins. Providing evidence for those implications was a top priority for Huxley. Nevertheless, Huxley continued to publish monographs on fossils and works on classification.[14] In 1867 he presented a major monograph on bird classification that set the stage for his classic paper "On the Animals Which Are Most Nearly Intermediate between Birds and Reptiles."[15]

In his opening remarks of "On the Classification of Birds," Huxley unequivocally asserted the close relationship between birds and reptiles. "The members of the class AVES so nearly approach the REPTILIA in all the essential and fundamental points of their structure, that the phrase 'Birds are greatly modified Reptiles' would hardly be an exaggerated expression of the closeness of that resemblance."[16] One could also say that reptiles were greatly modified birds because in reality they both were "different superstructures raised upon the one and the same ground plan."[17] But since some reptiles deviated very little from the general vertebrate plan, while all birds deviated significantly, Huxley believed that the reptilian features "might be taken to represent that which is common to both classes without any serious error."[18] Elsewhere, Huxley had proposed that birds and reptiles should be grouped together in a new category that he named the Sauropsida. Listing fourteen distinct points that demonstrated how they were similar to each other and yet distinct from mammals, he believed this justified them being grouped together.[19] Nevertheless, birds were also distinct from reptiles, and this monograph dealt only with classification within the Aves. Huxley realized that any taxonomy based on logical categories was more or less artificial. In a letter to the editor of *Ibis*, he wrote that such a classification was just the first stage in a process toward the ultimate goal, which was a "genetic classification—a classification which shall express the manner in which living beings have evolved one from the other."[20] The next stage was a classification based on gradation and formation of a natural series. He believed this should not be "confounded with the ultimate result because there was probably no way of knowing that absolutely." Nevertheless, such an ordering may "represent a true genetic classification more nearly than any other arrangement can do."[21] Combining fossil evidence with knowledge from comparative anatomy of modern birds, Huxley proposed a classification that was partly based on phylogeny. In this scheme *Archaeopteryx* was the only member of the order Saurorae and thus had not given rise to modern birds. Huxley's inconsistency is apparent in this discussion. On the one hand he believes that trying to base a classification on phylogeny is doomed to failure because the fossil record is so poor, but on the other hand, the ultimate natural classification for him was a phylogenetic one: "how living beings have evolved one from the other."

Archaeopteryx may not have captured his imagination in the early 1860s, but by the time Huxley presented his 1868 paper to the Royal Society, he was well aware of the evolutionary importance not only of *Archaeopteryx*, but also of a small bipedal reptile, *Compsognathus*, that had also been found at Solnhofen. In particular, he realized these fossils provided important evidence for gradual evolution because they represented transitional organisms across the boundaries of widely separated classes.

In spite of being an enthusiastic supporter of Darwin's theory, Huxley had always pointed out what he considered to be weaknesses in it. His monograph "On the Animals Which Are Most Nearly Intermediate between Birds and Reptiles" was no exception. He began the lecture by stating that although he firmly believed in the doctrine of evolution, that did not mean that he accepted the Darwinian hypothesis "in all its integrity and fullness." As he had done many times in the past, Huxley presented what had always been a highly problematic aspect of the theory—gradualism. He asked his audience, "How is it if all animals have proceeded by gradual modification from a common stock that these great gaps exist?"[22] Huxley acknowledged the fossil record was imperfect and that only a small part of the record had been examined. Nevertheless, he believed that this was an excuse. With all the fossils that had been discovered, surely there should be at least a few transitional organisms. With his usual flair for the dramatic, Huxley juxtaposed the hummingbird with the tortoise, the ostrich with the crocodile, and the stork with "the snake it swallows." How could these two groups be related? He listed seven distinct anatomical differences between present-day birds and reptiles, such as the position of the illium and ischia, the axis of the thigh bones, the number of toes, and ossification of the breast bone. Still, he admitted that transitional forms in many cases did exist and that the birds and reptiles were more closely related than superficial comparison let on. Drawing from his work on bird classification, Huxley pointed out that present-day birds of the family *Ratitae* such as the rhea and emu were more reptilian than other birds. Nevertheless, "the total amount of approximation to the reptilian type is but small, and the gap between Reptiles and Birds is but very slightly narrowed by their existence."[23] In the past, he would have continued this train of thought, arguing that since the gaps still remained, the evidence supported a saltational rather than a gradual evolutionary process. But the evidence had changed, and Huxley's lecture took a very different turn.

Huxley asked, "Are any fossil birds more reptilian than any of those now living? Are any fossil reptiles more bird-like than living reptiles?"[24] He answered yes to both questions. He described *Archaeopteryx* and demonstrated that it was much more reptilian in structure than any modern bird. Although *Archaeopteryx* appeared as an intermediate form between birds and reptiles, Huxley agreed with Owen's assessment that it was a bird, albeit with many reptilian features. If it was classified as a bird, it could not be a transitional form between reptiles and birds. A form such as *Archaeopteryx* that looked transitional but actually was not, Huxley later called an intercalary type. True transitional forms were forms that showed "a series of natural gradations between the reptiles and the bird, and enable us to understand the manner in which the reptilian has been metamorphosed into the bird type."[25] Such a type Huxley called a linear form.

If *Archaeopteryx* did not represent a linear type, what did? Huxley suggested that perhaps the pterodactyls or flying reptiles were the transitional form. They had air cavities in their bones; the coracoid, scapula, the broad sternum with its median all seemed "wonderfully bird like."[26] Some pterodactyls looked as if they had beaks sheathed in horn. However, Huxley did not think pterodactyls represented a linear form; instead they were another intercalary form. Most pterodactyls, except the giant ones, had teeth. More important, the vertebral column, hind limbs, and construction of the wing were quite different from those of birds. They were similar to birds in the ways that bats among mammals were similar to birds; in other words, they were "a sort of reptilian bat."[27] Therefore, they did not represent a link between reptiles and birds any more than a bat was a link between mammals and birds.

The missing link, the transitional or linear form, according to Huxley, was not a flying reptile, but rather would be found among the dinosaurs. Agreeing with the work of Hermann von Beyer, he classified dinosaurs as a "peculiar group of the *Reptilia*."[28] One particular lineage gave rise to birds. It was not just the weakness of the forelimb, but other more significant characteristics that showed dinosaurs were intermediate between reptiles and birds. Describing *Compsognathus longipes*, Huxley commented, "It is impossible to look at the conformation of this strange reptile and to doubt that it hopped or walked in an erect or semi-erect position, after the manner of a bird, to which its long neck, slight head and small anterior limbs must have given it an extraordinary resemblance."[29] Many of the tracks in the sandstones of Connecticut from the Triassic "are wholly undistinguishable from those of modern birds in form and size; others are gigantic three-toe impressions, like those of the Weald of our own country; others are more like the marks left by existing reptiles or Amphibia."[30] Huxley summarized his somewhat technical discussion:

> The important truth which these tracts reveal is that at the commencement of the Mesozoic epoch bipedal animals existed which had the feet of birds, and walked in the same erect or semi-erect fashion. These bipeds were either birds or reptiles or more probably both; and it can hardly be doubted that a lithographic slate of Triassic age would yield birds so much more reptilian than *Archaeopteryx*, and reptiles so much more ornithic than *Compsognathus*, as to obliterate completely the gap which they still leave between reptiles and birds.[31]

Huxley now accepted that Darwin was correct in claiming that the incompleteness of the fossil record was responsible for the lack of transitional forms. As incomplete as the record was, evidence was coming in that was starting to fill the gaps. The evidence was still quite fragmentary, and Huxley did not think that *Compsognathus* was the true missing link between birds and reptiles. Huxley believed that "many completely differentiated birds in all probability existed even in the Triassic epoch, and as we possess hardly any knowledge of the terrestrial reptiles of that period, it may be regarded as certain that we have no knowledge of the animals which linked Reptiles and Birds together historically and genetically."[32]

Compsognathus and *Archaeopteryx* may not have been the true missing link between birds and reptiles, but they were crucial evidence in support of gradualism because they showed that "in former periods of the world's history there were animals which overstepped the bounds of existing groups, and tend to merge them into larger assemblages."[33] Nevertheless, in Huxley's second lecture to the Americans, he admitted that this evidence "does not prove that birds have originated from reptiles by the gradual modification of the ordinary reptile into a dinosaurian form, and so into a bird," only that such a process "may possibly have taken place."[34] This evidence was very strongly in favor of evolution, but it was not the highest kind of evidence. For Huxley, the "demonstrative evidence of evolution" did not yet exist for the reptile-bird transition. To be truly demonstrative of evolution one needed to find a series of forms that progressed in a steplike manner showing the metamorphosis of reptile to bird. "In some ancient formation reptiles alone should be found; in some later formations birds should first be met with, and in the intermediate strata we should discover in regular succession those forms which I pointed out to you which are intermediate between reptiles and birds."[35] While many bird-reptile transitional forms existed, they were found in contemporaneous deposits. "They do not occur in the exact order in which they ought to occur, if they really had formed steps in the progression from reptile to the bird."[36] However, this type of evidence had recently been "forthcoming in considerable and continually increasing quantity . . . in many divisions of the animal kingdom."[37] For Huxley, this type of evidence completed the proof of evolution.

Equus—The Demonstrative Evidence of Evolution

In Huxley's final lecture on evolution to the Americans, he presented what he considered to be the highest form of evidence in favor of evolution—the pedigree of the horse. In a brilliant rhetorical move, Huxley described the horse using the language of the natural theologians:

> The horse . . . presents us with an example of one of the most perfect species of machinery in the animal kingdom. . . . Among mammals it cannot be said that there is any locomotive so perfectly adapted to its purposes, doing so much work with so small a quantity of fuel, as this animal. . . . [T]he horse is a beautiful creature. . . . Look at the perfect balance of its form, and the rhythm and perfection of its action. The locomotive apparatus . . . its slender fore and hind limbs . . . are flexible and elastic levers, capable of being moved by very powerful muscles; and in order to supply the engines which work these levers with the force which they expend, the horse is provided with a very perfect feeding apparatus, a very perfect digestive apparatus.[38]

And yet, Huxley would demonstrate that this perfectly adapted creature came into existence by a process of gradual evolution, rather than as a result of Divine Plan. But to do so, Huxley had to take his listeners through a somewhat technical discussion of the anatomy of the horse—particularly the structure of its limbs.

With the aid of diagrams, Huxley compared the bones in the limbs of the horse to human bones, showing which ones corresponded and how they were modified such that they had been "converted into long, solid springy, elastic levers, which are the great instruments of locomotion of the horse."[39] This strategy served two purposes. First, it allowed lay people to follow a somewhat complicated discussion about horse bones by relating it to the bones in something they were more familiar with—themselves. More importantly, it illustrated that humans and horses might not be as different as they thought, but were modifications of a general mammalian type. The characteristics of the general mammalian type included "the possession of a perfectly distinct radius and ulna, two separate and distinct movable bones. . . . [M]ammals in general possess five toes, often unequal, but still as completely developed as the five digits of my hand."[40] The general type also had a complete tibia and fibula, which were distinct separable bones, and hind feet that possessed five distinct toes. Thus, a differentiated animal such as the horse must have "proceeded by way of evolution or gradual modification from a form possessing all the characteristics we find in mammals in general."[41] If evolution was true, then somewhere in the rocks should be preserved a series of fossils that would show the various stages through which the horse had passed. "Those stages ought gradually to lead us back to some sort of animal which possessed a radius, and an ulna, and distinct complete tibia, and fibula, and in which there were five toes upon the fore-limb, no less than upon the hind-limb."[42]

In addition to the skeleton, Huxley also discussed the teeth of the horse. Most horses had no more than forty teeth, many only thirty-six. This was because while the males had canine teeth, the "tushes or canine teeth of the mare are rarely developed."[43] Up to a certain age a dark mark existed in the incisors of the horse. In young foals, the incisors had a very deep pit and then a long fang. But as it aged, the pit filled up with fodder, which became carbonized, leaving a dark mark. As the horse grew older, the tooth was worn down, even beyond the bottom of the pit, and thus the mark disappeared. The horse also had a series of grinding teeth that were quite large with extremely long crowns, which took a long time to wear down. The crowns, made of enamel, dentine, and a soft bony matter called cement, exhibited a very complicated pattern of ridges and crescents. The different degrees of hardness of the various materials caused uneven wear with the resultant pattern of ridges.[44] The practical effect of such a construction Huxley compared to the lamination of a millstone. The lamination meant that "the ridges wear less swiftly than the intermediate substance, and therefore the surface always keeps rough and exerts a crushing effect upon the grain."[45] The same was true of the horse tooth; the surface was kept irregular, and "that has a very great influence upon the rapid mastication of the hard grain or the hay upon which the horse subsists."[46] How did this compare to the general mammalian type tooth structure? Most higher mammals had approximately forty-four teeth. The incisors had no pit, and the grinding teeth as a rule increased in size from the front toward the back part of the series. Thus, if the theory of evolution was correct, Huxley claimed that the series of fossil forms that preceded the horse should show

> animals in which the mark upon the incisor gradually more and more disappears, animals in which the canine teeth are present in both sexes and animals in which the teeth gradually lose the complications of their crowns and have a simpler and shorter crown, while at the same time they gradually increase in size from the anterior end of the series towards the posterior.[47]

Having suggested what should be found in the fossil record according to the hypothesis of evolution, Huxley invited his audience to "turn to the facts and see how they bear upon the requirements of this doctrine of evolution."[48]

Huxley devoted his third lecture to the genealogy of the

horse, but the lecture he gave was very different from the one that he had originally planned, which had been based entirely on European fossils. The American paleontologist Othniel C. Marsh had also been working on the genealogy of the horse. However, the fossils from North America had led him to quite different conclusions about the origins of the horse. Marsh believed that his specimens showed conclusively that the horse had originated in the new world and not in the old, and thus its genealogy must be worked out in the new world. Somewhat hesitantly, Marsh put the matter before Huxley, who spent nearly two days going over the specimens, testing each point that Marsh had made.[49]

At each inquiry, whether he had a specimen to illustrate such and such a point or exemplify a transition from earlier and less specialized forms to later and more specialized ones, Professor Marsh would simply turn to his assistant and bid him fetch box number so and so, until Huxley turned upon him and said, "I believe you are a magician; whatever I want you just conjure it up."[50]

The American fossils represented all new information to Huxley, and he told Marsh that these facts "demonstrated the evolution of the horse beyond question, and for the first time indicated the direct line of descent of an existing animal."[51] Huxley gave up his own views, agreeing with Marsh that the European fossils were the result of occasional immigrants from the new world. He frantically prepared virtually an entirely new lecture on the genealogy of the horse to include the paleontological finds of Marsh for his third and final lecture on evolution.

Beginning with fossils from the most recent period, Huxley described a series of fossils related to the horse. Europe and England had many fossil horses from the Pliocene that had a structure virtually identical to existing horses. The earlier Pliocene and later Miocene deposits in Germany, Greece, India, Britain, and France also contained fossils that were "so entirely like that of the horse that you may follow descriptions given in works upon the anatomy of the horse upon the skeletons of these animals."[52] Named *Hipparion*, it had certain significant differences. Attached to the long metacarpal bone were two splintlike bones, and attached to the end of them was a small toe with three joints of the same general character as the middle toe. But they were very much smaller and must have been essentially nonfunctional, similar to the dew-claws (vestigial digits) found in many ruminant animals. Nevertheless, these lateral toes, which were "almost abortive in the existing horse," were fully developed in *Hipparion*.[53] The hind limb had a fibula that was pretty much the same as the modern horse. However, the fore limb had a distinct ulna. It was very thin but was "traceable down to the extremity."[54] Huxley then turned to the middle and older parts of the Miocene to describe a series of fossils that he referred to as making up the genus *Anchitherium*. They were still clearly like the horse in their general organization, but in these forms, all four feet had three distinct functional toes, not the dew-claws of *Hipparion*. The ulna of the forearm was quite distinct from the radius along its whole length, although closely united with it. The hind leg had a similar structure to the fore leg, but with a better developed fibula.

The most significant change, however, was in the teeth. There was only a rudimentary pit in the incisors, and the canines were present in both sexes. The molars were short with no cement. With the aid of diagrams, Huxley demonstrated that the pattern was simply a less complex form of the one found in modern horse teeth. He admitted that there was no way to know whether these species represented the exact line of modification leading to the modern horse. Nevertheless, these three forms represented a succession of the horse type, with the oldest being closest to the general mammalian type. Through time, these forms had "undergone a reduction of the number of their toes, a reduction of the fibula, a more complete coalescence of the ulna with the radius. The pattern of the molar teeth had become more complicated and the interspaces of

their ridges had become filled with cement."[55] Huxley concluded, "In this succession of forms you have exactly that which
the hypothesis of evolution demands. The history corresponds
exactly with that which you would construct *a priori* from the
principles of evolution."[56] He was not referring to the principle
of natural selection but rather meant that forms would become
better adapted to their environment, becoming more specialized over time. Attacking the alternative explanation that stated
each form had been separately created at separate epochs of
time, Huxley again reminded his listeners that there could
never be scientific evidence for such a hypothesis, and "it is not
pretended that there is the slightest evidence of any other kind
that such successive creation has ever taken place."[57]

Huxley then turned to the fossil finds of Marsh. To the great
pleasure of his American audience, he claimed that "there is
nothing in any way comparable, for extent, or for the care with
which the remains have been got together or for their scientific
importance to the series of fossils which he has brought
together."[58] Huxley described the American fossils that corresponded very closely to the European ones, but with a few significant differences. Concurring with Marsh, he believed that
these differences meant that the European *Hipparion* was a side
branch rather than in the direct line of succession to the modern
horse. However, the American fossils were important for
another reason. They included a series of forms that were far
older than any that had been found in Europe. The lower
Miocene form *Mesohippus* had three toes in front and a large
splintlike rudiment representing the little finger. The radius and
ulna were entire bones, and the tibia and fibula were also distinct. The teeth were anchitoeroid with short crowns. But for
Huxley the most important find was *Orohippus* from the lower
part of the Eocene. It had four complete toes on the front limb,
three on the hind, a well-developed ulna, a well-developed
fibula, and the teeth of the simple pattern. Again, Huxley
pointed out that these forms were "exactly and precisely that

which could have been predicted from a knowledge of the principles of evolution."[59] In fact, Huxley predicted that equine fossils from the Cretaceous would have four complete toes with a rudiment of the innermost toe in front and probably a rudiment of the fourth toe in the hind foot. And sure enough, two months after Huxley gave this lecture, Marsh discovered such a form from the lowest Eocene deposits of the American West.[60]

Huxley concluded his final lecture by returning to the underlying theme of the whole series: What was a scientific hypothesis and how was it proved? "An inductive hypothesis is said to be demonstrated when the facts are shown to be in entire accordance with it. If that is not scientific proof, there are no inductive conclusions which can be said to be scientific."[61] Huxley had presented several kinds of evidence. Some, such as persistent types, he admitted did not demonstrate evolution but were consistent with the theory. The reptile-bird transitional fossils provided evidence strongly in favor of evolution, while the series of fossil horses was demonstrative proof of evolution. For Huxley "the doctrine of evolution at the present time rests upon exactly as secure a foundation as the Copernican theory of the motions of the heavenly bodies. Its basis is precisely of the same character— the coincidence of the observed facts with theoretical requirements."[62] With one final dig at the creationists, he maintained that there was no scientific evidence nor could there be any in support of the supposition that the forms had been created separately at separate epochs. Furthermore, he knew of no other hypotheses which were supported or pretended to be supported by evidence or authority of any other kind. He only hoped that

> the time will come when such suggestions as these, such obvious attempts to escape the force of demonstration, will be put upon the same footing as the supposition made by some writers, who are, I believe, not completely extinct at present, that fossils are not real existences, are no indications of the existence of the animals to which they seem to belong; but that

they are either sports of Nature or special creations, intended—
as I heard suggested the other day—to test our faith.[63]

Of course, such an idea was preposterous to Huxley. One
had to simply examine the facts, and in the case of evolution,
the facts were unequivocal. "The whole evidence is in favor of
evolution, and there is none against it."[64]

Was Gradualism Proof of Evolution?

The discovery of the birdlike dinosaurs and *Archaeopteryx* were
key factors in Huxley's abandoning saltation. The series of
horse fossils provided further evidence for gradualism. But in
discussing these fossils as support for gradualism, I have simultaneously presented them as evidence that Huxley considered
definitive proof of the general theory of evolution. Does this
mean that Huxley did not really accept evolution until 1867–
1868 when he was studying the *Dinosauria*? I do not think so.
Huxley would not have championed a hypothesis that he did
not think was true overall, even if he disagreed with the details.
Just as in the recent debates over gradualism versus punctuated
equilibria, one can argue over the relative frequencies of the different modes of speciation, but this does not mean that one is
questioning whether evolution actually occurred. However,
Huxley's changing views about gradualism highlights his attitude toward scientific theories and why Darwinism captured his
attention in a way that Lamarck's theory had not.

I pointed out that an underlying theme of Huxley's American lectures was to use evolution to explain what constituted a
scientific hypothesis. In contrasting Darwin's theory with other
theories about the history of life, Huxley specifically avoided
the word "creation," but instead described the "Miltonic hypothesis." Unlike the creation hypothesis, whose support ultimately relied on faith, the Miltonic hypothesis could be tested

empirically. The Darwinian hypothesis and the Miltonic hypothesis could be compared just on the "facts." It is true that structuring the discussion this way served his larger purpose of keeping science separate from theology, and it is also true that Huxley certainly was naive in his belief that what constituted a "fact" could be clearly identified. Nevertheless, Darwin's theory made sense of a great number of "facts" of nature. And the more speculative aspects of the theory, such as gradualism, could be subjected to scientific evaluation. Did the evidence support gradualism? In 1859 Huxley answered no. This did not prevent Huxley from supporting Darwin's theory, although he realized that gradualism was an intrinsic part of it. Thus, if a series of fossils could be found that showed a gradual transition from one form to another, this would be powerful support for evolution. But Huxley did not wish to conflate gradualism with evolution. Evolution was a theory about the history of life. Gradualism was a description of the pattern of change that could be evaluated by examining the fossil record. Always the empiricist, Huxley wanted evidence he could see. As will be explained more fully in chapter 7, lack of evidence was also the basis of his doubting the efficacy of natural selection. Artificial selection had not yet succeeded in producing a new species from varieties. While Huxley eventually accepted gradualism, he maintained that his saltational views of the early 1860s better fit the fossil evidence at that time.

Following the publication of the *Origin*, research in many different fields provided evidence for Darwin's gradualistic scheme. Paleontological discoveries and research in development and comparative anatomy made the idea of gradual change more and more plausible. Huxley was well aware of this research; indeed, he was an active participant in it, and one should not underestimate its influence on his change of opinion. This is clearly seen in his 1880 retrospective "On the Coming of Age of the *Origin of Species*," in which he discussed the changes in opinion that had occurred in the twenty-one years since the

Origin had been published. Huxley was not alone in his belief that very little evidence existed in 1859 to support Darwin's claim that all organisms must have at one time been connected by direct or indirect intermediate gradations.

No part of Mr. Darwin's work ran more directly counter to the prepossessions of naturalists twenty years ago than this. And such prepossessions were very excusable, for there was undoubtedly a great deal to be said, at that time, in favour of the fixity of species and of the existence of great breaks, which there was no obvious or probable means of filling up, between various groups of organic beings.[65]

The structural breaks between certain present-day organisms were real, and Darwin could account for them only by claiming that the intermediate forms had become extinct. Huxley certainly did not deny the prevalence of extinction. Nevertheless, the breaks seemed too large and too many for this to be a complete explanation. But Huxley admitted that the "progress of knowledge has justified Mr. Darwin to an extent which could hardly have been anticipated."[66] Huxley summarized the research that had been done since 1859 that all provided more and more evidence in favor of gradualism. The discovery of *Archaeopteryx* and his own work on the dinosaurs were key.

Since 1868 more evidence had come to light supporting the idea that dinosaurs were the common ancestor to birds and reptiles. The structure of the pelvis as shown by *Megalosaurus*, *Iguanodon*, and *Hypsilophodon*, and the distal end of the tibia and the astragalus in *Poikilopleuron*, *Megalosaurus*, and *Laelaps* all showed "the ornithic affinities of the dinosauria."[67] Huxley had pointed out in his 1868 monograph that the reptiles and birds were clearly distinct with respect to the bones known as the ilium and the ischia. But the dinosaurs were intermediate between the two. "The ischia of a Dinosaurian are more bird like than those of any existing reptile, but retain the reptilian union in the symphysis."[68] *Hypsilophodon*, in particular, "affords unequivocal evidence of a further step towards the bird. . . . If only the pubis and the ischium of *Hypsilophodon* had been discovered, they would have unhesitatingly referred to Aves."[69] However, not all the dinosaurs showed these modifications. The *Dinosauria* were evidence in favor of gradualism because they "present us with serial modifications leading from the Parasuchian type of structure on the one hand, to that of Birds on the other."[70] Huxley concluded his highly technical discussion of the transitional nature of the *Dinosauria* by describing the English bird the dorking fowl. Certain bones, if found in a fossil state, would be indistinguishable from the bones of a dinosaur. And if it were somehow possible to enlarge and fossilize the hindquarters of a half-hatched chicken, "they would furnish us with the last step of the transition between Birds and Reptiles; for there would be nothing in their characters to prevent us from referring them to the *Dinosauria*."[71]

The remains of *Archaeopteryx* indicated that it was a very primitive, reptilianlike bird. One characteristic that distinguished birds from reptiles was the absence of teeth. The head of *Archaeopteryx* was missing, so no connections between birds and reptiles could be drawn on that front. Thus, another important piece of evidence in favor of the reptilian origins of birds would be the discovery of toothed birds. In the Cretaceous deposits of the United States, Marsh had discovered such fossils. Both *Hesperonis* and *Ichthyornis* had true teeth like those of a reptile—not just serrations of the jaw that were seen in many birds.[72] The bodies of the vertebra in *Ichthyornis* were biconcave (concave on both sides, as a lens) like those of the lower reptiles, rather than the saddle shape found in modern birds. "In tracing birds back in time we find a parallel series of modifications to those described in the Crocodilia."[73] The new evidence caused Huxley to claim that "the evolution of birds from reptiles . . . is by no means such a wild speculation as it might from *a priori* considerations have been supposed to be."[74]

The fossil record was providing links between different classes and documenting serial modifications of particular lineages. Perhaps even more dramatic evidence in favor of the common ancestry of all organisms was the research in development and comparative anatomy. In 1859 "there appeared to be a very sharp and clear hiatus between vertebrated and invertebrated animals, not only in their structure, but what was more important, in their development."[75] Huxley admitted that the exact details of the connection between the two groups were still not known. Nevertheless, the investigations of the Russian embryologist Aleksandr Kowalesky and others on the development of *Amphioxus* and of the *Tunicata* "prove, beyond a doubt that the differences which were supposed to constitute a barrier between the two are non-existent."[76] Another complete separation appeared to exist between flowering and flowerless plants. But the investigations of the German botanist Wilhelm Hofmeister on the modifications of the reproductive apparatus in the *Lycopdiaceoe*, the *Rhizocarpeoe*, and the *Gymnospermeoe* had shown how "the ferns and the mosses are gradually connected with the Phaneorganic division of the vegetable world."[77] Finally, a great deal of knowledge had accumulated on the lowest forms of life, "which demonstrates the futility of any attempt to separate the lowest plants from the lowest animals, and show that the two kingdoms of living nature have a common borderland which belong to both, or to neither."[78] Indeed, in a separate technical article specifically on the relationship of plants to animals, Huxley demonstrated that "the difference between animal and plant is one of degree rather than of kind; and the problem whether, in a given case, an organism is an animal or a plant may be essentially insoluble."[79] Since 1859 the whole tendency of biological investigation "has been in the direction of removing the difficulties which the apparent breaks in the series created at that time."[80]

For Huxley, no matter how logical a theory seemed, if it was not backed by evidence, he remained skeptical. Darwin argued that the connecting links between major groups would be found among extinct organisms, but that we would have little evidence of such transitional organisms because the vast majority of them would not become fossilized. But Huxley was not convinced. He drew an analogy between a scientific doctrine and the lands of an estate:

> If a landed proprietor is asked to produce the title-deeds of his estate, and is obliged to reply that some of them were destroyed in a fire a century ago, and that some were carried off by a dishonest attorney; and that the rest are in a safe somewhere, but that he really cannot lay his hands upon them; he cannot I think, feel pleasantly secure, though all his allegations may be correct and his ownership indisputable.[81]

A scientific doctrine was like an estate, and the holder must be able to "produce his title deeds in the way of direct evidence" or take the consequences.[82] In 1859, Huxley believed that Darwin was in the same position as the landed proprietor of his example. By 1868, however, the situation had changed. Huxley wrote that "[i]f I cannot produce the complete title-deeds of the doctrine of animal Evolution, I am able to show a considerable piece of parchment evidently belonging to them."[83] *Archaeopteryx* and the birdlike dinosaurs were Huxley's "piece of parchment." "[T]hey enable us to form a conception of the manner in which Birds may have evolved from Reptiles, and therefore justify us maintaining the superiority of that hypothesis, that birds have been so originated to all hypotheses which *are devoid of an equivalent basis of fact*."[84] However, the dinosaurs were not the only evidence in favor of evolution. As the previous discussion indicates, after 1868 more and more pieces of parchment continued to turn up, showing that "Mr. Darwin did not over-estimate the imperfection of the geological record."[85] The most dramatic proof of this was the discovery of a series of fossil mammals. In addition to the fossil horses discovered by Marsh in America, the French paleontologist Albert Gaudry had found in the Pikermi deposits in Attica a series of fossils showing the transition from the ancient civets to modern hyenas. Among the phosphorites of Quercy, Filhol had described seventeen varieties of the genus *Cynodictis* that filled up the gap between the viverine animals such as civets and mongooses and the bearlike dog *Amphicyon*. The result of these investigations "has been to introduce to us a multitude of extinct animals, the existence of which was previously hardly suspected." For Huxley "evolution is no longer a speculation, but a statement of historical fact."[86]

Although Huxley initially disagreed with Darwin's emphasis on gradual change, as evidence accumulated demonstrating that some groups showed a series of gradations, he was more than willing to abandon his saltational views. The fossil record was still woefully incomplete. And Huxley was careful to point out that the particular fossils he used as examples, for instance in his discussion of the evolution of the horse, probably did not represent the exact lineage, but they demonstrated that a series of gradual changes were possible. Moreover, considering the fragmentary nature of the fossil record, they were probably as close to a phylogenetic series as would be possible. The series of horse fossils demonstrated unequivocally that evolution in the horse lineage had proceeded from a form that was a relatively generalized mammalian type to a form that was highly specialized and well adapted to its particular mode of life. Incomplete as the fossil record was, Huxley made use of it for Darwin's cause. Unlike Lyell, he grasped the significance of *Archaeopteryx* as a transitional organism. For Lyell, *Archaeopteryx* just meant that birds existed much earlier than had been suspected. Concluding his discussion about the "valuable relic," he wrote, "We may learn . . . how rashly the existence of Birds at the epoch of the Secondary rocks had been questioned, simply on negative evidence."[87] Throughout the *Antiquity of Man*, Lyell emphasized the fragmentary nature of the record. Instead of claiming that the fossil record was becoming better known and less imperfect, he stressed how much remained to be discovered. Martin Rudwick even claims that Lyell got himself into the peculiar position of believing the fossil record *could not* provide evidence for evolution. "Lyell failed to suggest that a search might be made for *successions* of fossil species occurring in an order that would be meaningful in evolutionary terms, even if the succession was too incomplete to show trans-specific changes in detail."[88] I think Rudwick overstates his case. I never found anything in Lyell's writings that explicitly denied the possibility of using paleontological data in support of Darwin's theory. Nevertheless, he also did not use fossils to argue for Darwinism. But Huxley heralded the many new fossil discoveries as "demonstrable proof" of evolution.

Previously, I argued that Huxley denied progression in part because he associated it with the argument from design. However, the evidence for progression was also quite contradictory. Huxley could and did make a plausible case against progression based purely on fossil evidence. Although his antitheological views may have been responsible for his denial of progression, ultimately the "facts" carried the day. With all the fossils that had been discovered, vertebrates still appeared later than invertebrates, mammals later than fish. As in the case of gradualism, when the evidence changed, so did Huxley's views. He was willing to accept that the fossil record was progressive, in the most general sense of the word. I also claimed that in his zeal to separate science from theology, he denied some of the excellent work done by paleontologists in the service of natural theology. For the most part, such an attitude served Huxley well and meant that his scientific views were primarily shaped by the evidence. While this is supposed to be true of all good scientists, most people today acknowledge that cultural, political, and personal factors often significantly influence the practice of science. Specifically, the role that personal relationships between scientists play in scientific debates should not be underestimated. Lyell was a particularly important influence on Huxley in shaping his views on both saltation and progression. However, Huxley differed from Lyell in significant ways. While it was Lyell's religious beliefs that prevented him from ever fully accepting evolution, Huxley's quarrel with various aspects of Darwin's theory was firmly grounded in what he considered lack of adequate empirical evidence. The next chapter examines Huxley's role in the debates over the implications of evolutionary theory for human origins. Once again, Huxley's desire to keep theology out of the discussion is an underlying theme as well as a strategy to convince. Personality plays a prominent role in this debate in Huxley's clash with Richard Owen. But most importantly, Huxley grounds his argument for the pithecoid ancestry of humans in facts—cold, irrefutable facts of comparative anatomy.

Notes

1. Thomas Huxley, 1860, "The Origin of Species," in *Collected Essays of Thomas Huxley*, vol. 2, *Darwiniana* (London: Macmillan & Co., 1893), p. 77.

2. Thomas Huxley, "On the Recent Work of the 'Challenger' Expedition, and Its Bearing on Geological Problems," notes in *Proceedings of the Royal Institute of Great Britain* 7 (1875): 354–57, *Scientific Memoirs of Thomas Henry Huxley* (*SMTHH*), 4 vols., Michael Foster and E. Ray Lancaster, eds. (London: Macmillan & Co., 1898–1902), 4: 64.

3. Thomas Huxley, November 23, 1859, *Life and Letters of T. H. Huxley* (*LLTHH*), 2 vols., Leonard Huxley, ed. (New York: D. Appleton & Co., 1900), 1: 189.

4. Charles Darwin, *The Origin of Species* (1859; New York: Avenel, 1976), p. 413.

5. Thomas Huxley, "Autobiography," in *Collected Essays of Thomas Huxley*, vol. 1, *Methods and Results* (London: Macmillan & Co., 1893), p. 15.

6. Thomas Huxley, 1880, "The Coming of Age of the Origin of Species," *Darwiniana*, p. 241.

7. Mario di Gregorio, *T. H. Huxley's Place in Natural Science* (New Haven, Conn.: Yale University Press, 1984), pp. 77–78.

8. Again, Gregorio disputes this, claiming that Huxley wrote *Man's*

Place in Nature to discredit Owen (ibid., p. 152). I shall address Gregorio's charge in the next chapter.

9. Thomas Huxley, January 21, 1868, *LLTHH* 1: 325.

10. Ernst Haeckel, *Generelle Morphologie der Organismen*. Quoted in di Gregorio, *T. H. Huxley's Place*, pp. 79–80.

11. Richard Owen, "On the Fossil Remains of a Longtailed Bird (*Archaeopteryx macrurus*) from the Lithographic Slate of Solenhofen," *Proceedings of the Royal Society of London* 12 (1862): 273.

12. Ibid.

13. Adrian Desmond, *Archetypes and Ancestors* (Chicago: University of Chicago Press, 1984), p. 128.

14. Thomas Huxley, "On the Classification of Birds; and on the Taxonomic Value of the Modifications of Certain of the Cranial Bones Observed in that Class," *Proceedings of the Zoology Society of London* (1867): 415–72, *SMTHH* 3: 238–97.

15. Thomas Huxley, February 7, 1868, "On the Animals Which Are Most Nearly Intermediate between Birds and Reptiles," *Geology Magazine* 5 (1868): 357–65, *SMTHH* 3: 303–13.

16. Huxley, "On the Classification of Birds," *SMTHH* 3: 238.

17. Ibid.

18. Ibid.

19. See Thomas Huxley, 1865, "Explanatory Preface to the Catalogue of the Paleontological Collection in the Museum of Practical Geology," *SMTHH* 3: 125–75.

20. Huxley, "On the Classification of Birds," *SMTHH* 3: 296.

21. Ibid.

22. Huxley, "On the Animals," *SMTHH* 3: 304.

23. Ibid., p. 306.

24. Ibid., p. 307.

25. Thomas Huxley, 1876, "The Hypothesis of Evolution: The Neutral and the Favorable Evidence," *American Addresses with a Lecture on the Study of Biology* (London: Macmillan and Co., 1877), p. 59.

26. Huxley, "On the Animals," *SMTHH* 3: 308.

27. Ibid.

28. Thomas Huxley, 1870, "On the Classification of the Dinosauria with Observations on the Dinosauria of the Trias," *SMTHH* 3: 487.

29. Huxley, "On the Animals," *SMTHH* 3: 311–12.

30. Ibid., p. 312.

31. Ibid.

32. Ibid., pp. 312–13.

33. Huxley, *American Addresses*, p. 58.

34. Thomas Huxley, "The Negative and Favorable Evidence," *Popular Science Monthly* 10 (1877): 223. Huxley's American lectures appeared both in the book *American Addresses* and in *Popular Science Monthly*. The *Popular Science Monthly* version was based on the report of the *New York Tribune* but had been revised by Huxley. The two versions are not identical. For instance, there was no use of the terms "linear" and "intercalary" in the *Popular Science* version. Therefore, I have used both sources. But both sources are referring to a series of three lectures that he gave at Chickering Hall, New York, on September 16, 18, and 20, 1876.

35. Thomas Huxley, "The Demonstrative Evidence of Evolution," *Popular Science Monthly* 10 (1877): 286.

36. Ibid.

37. Ibid.

38. Ibid., p. 287.

39. Ibid., p. 289.

40. Ibid., p. 290.

41. Ibid.

42. Ibid.

43. Ibid., p. 289.

44. Ibid.

45. Ibid., p. 290.

46. Ibid.

47. Ibid., pp. 290–91.

48. Ibid., p. 291.

49. Othniel C. Marsh, August 1895, *LLTHH* 1: 495.

50. Leonard Huxley, 1900, *LLTHH* 1: 495.

51. Marsh describing Huxley's reaction, August 1895, *American Journal of Science* 1. Quoted in *LLTHH* 1: 495.

52. Thomas Huxley, "The Demonstrative Evidence of Evolution," *Popular Science Monthly* 10 (1877): 291.

53. Ibid.

54. Ibid.

55. Ibid., p. 292.

56. Ibid.

57. Ibid., p. 293.

58. Ibid.

59. Ibid., p. 294.

60. Ibid., p. 296(n). See also *American Journal of Science*, November 1876.

61. Ibid.

62. Ibid.

63. Ibid.

64. Ibid.

65. Thomas Huxley, 1880, "On the Coming of Age of *The Origin of Species*," *Darwiniana*, p. 233.

66. Ibid., p. 234.

67. Thomas Huxley, "Further Evidence of the Affinity Between the Dinosaurian Reptiles and Birds," *Quarterly Journal Geology Society London* 12 (1870): 12–31, *SMTHH* 3: 480.

68. Ibid., p. 481.

69. Ibid., p. 482.

70. Ibid. "Parasuchian" referred to a reptile from the Indian Trias allied to *Belodon* about which Huxley was planning to describe in a separate monograph.

71. Ibid., p. 486.

72. Thomas Huxley, "On the Evidence as to the Origin of Existing Vertebrate Animals," *Nature* 13 (1876): 388–89ff., *SMTHH* 4: 180.

73. Ibid.

74. Ibid., p. 182.

75. Huxley, "On the Coming of Age," p. 235.

76. Ibid., p. 236.

77. Ibid.

78. Ibid.

79. Thomas Huxley, "On the Border Territory Between the Animal and the Vegetable Kingdoms," *Macmillan's Magazine* 33 (1876): 373–84, *SMTHH* 4: 162.

80. Huxley, "On the Coming of Age," p. 236.

81. Huxley, "On the Animals," *SMTHH* 2: 303.

82. Ibid., p. 305.

83. Ibid.

84. Ibid., p. 313 (emphasis added).

85. Huxley, "On the Coming of Age," p. 238.

86. Ibid., pp. 238–39, 242.

87. Charles Lyell, *Geological Evidences of the Antiquity of Man* (Philadelphia: George W. Childs, 1863), p. 453.

88. Martin Rudwick, *The Meaning of Fossils* (Chicago: University of Chicago Press, 1982), p. 244.

Chapter Six

Convincing Men They Are Monkeys

By next Friday evening they will all be convinced that they are monkeys.[1]

H uxley has been characterized as "Darwin's bulldog," suggesting that he primarily popularized and defended Darwin's theory rather than contributed original research to support evolution. But this was clearly not the case. As the previous chapter demonstrates, Huxley's research on *Archaeopteryx*, the dinosaurs, and on the ancestry of the horse provided important evidence for evolution. Furthermore, Huxley aided Darwin's cause many years earlier. Although Darwin clearly thought humans were a part of Nature's evolutionary scheme, the *Origin* did not directly discuss the evolution of man. Darwin's cryptic comment on the matter was "Light will be thrown on the origin of man and his history."[2] Huxley immediately saw the light. *Man's Place in Nature*, published in 1863, eight years before Darwin published his *Descent of Man*, shows that Huxley fully grasped the implications of evolution for human origins and that he did not balk at those implications. *Man's Place in Nature* argued eloquently and powerfully that humans were no exception to the theory of evolution.

While many aspects of this story are well known, it is worthwhile to examine exactly how Huxley chose to defend Darwin for a variety of reasons—not the least of which is that Huxley correctly perceived the most controversial aspect of Darwinism: the question of human ancestry. Thus, this is where he initially devoted most of his energy. While scientists argued whether natural selection could account for species change, whether change was saltational or gradual, and a myriad of other aspects of Darwin's theory, this was not the fundamental quarrel with the *Origin*. Opponents of Darwin's theory attacked his book, claiming it was materialistic, atheistic, and worse. Darwin was telling Victorians that rather than being made in God's image, humans were little more than intelligent apes.

Huxley was not merely interested in convincing "men that they were monkeys." Examining Huxley's promulgation of evolution, it becomes apparent that he had a more ambitious agenda than just defending Darwin. He used evolution theory to attack theology and to further his broad campaign to replace the power and moral authority of the Church with that of the temple of science. If in the process he also managed to make fools of his enemies, this was the icing to the proverbial cake. In the controversy surrounding the ape-human question, science, politics, and personality conflicts were inextricably linked. Furthermore, the fracas spilled beyond the hallowed halls of science, providing an entertaining and illuminating episode in the acceptance of evolutionary theory.

Apes and Humans Pre-Origin

While the publication of the *Origin* raised the problem of human origins to new heights of controversy, people had long been fascinated with the humanlike apes. The relationship of humans to apes and of apes to each other had become an issue as soon as such animals were discovered. Explorers greatly enjoyed the antics of what later were identified as chimpanzees and marveled at their ability to imitate the gestures of humans. However, the early descriptions were a mixture of fact and fantasy, and it was difficult to separate the two. Although cadavers of chimpanzees and orangutans had been dissected in the seventeenth century, the two species were often confused with each other as well as with the gibbon. The gorilla was not identified until the nineteenth century, and "orang-utang" was often used as a generic term for all nonhuman primates well into the nineteenth century. Furthermore, the exact relationship of the "orang-utang" to humans was problematic. Edward Tyson in an excellent memoir entitled *Orang-Outang: Or the Anatomy of a Pygmie Compared with That of a Monkey, an Ape, and a Man*, published by the Royal Society in 1699, detailed the remarkable similarities between the pygmy (clearly a chimpanzee according to Tyson's drawings and description) and humans. But he also recognized the differences: "[though it] does so much resemble a Man in many of its parts more than any of the ape kind or any other animal in the world that I know of; yet by no means do I look upon it as the product of a mixt generation—'tis a Brute animal sui generis and a particular species of Ape"[3] As Tyson pointed out in the dedication, the purpose of the monograph was to demonstrate the scale of nature in which the pygmy was the connecting link between the animal and the rational.[4] In his *System of Nature*, Linnaeus classified humans as quadrupeds, putting them in the same order as apes and sloths. Under the generic name *Simia satyrus* he placed the chimpanzees that had been described by Tulp, Edwards, and Scotin. But another group that had been described by Gesner, Johnston, and Brisson he gave the name *Simia sylvanus*. Because the descriptions he had gotten about the Borneo orang-outang seemed sufficiently different, he put this creature in a separate human genus, *Troglodytes*. Thus, a great deal of confusion existed with regard to the classification of these primates. In the genus *Homo* Linnaeus included not only our own species but several others that, in fact, did not exist. He had never seen these creatures firsthand, and his classification represented an amalgamation of traits based on old woodcuts and on a dissertation of his pupil Hoppius.[5] Georges Louis Leclerc, Comte de Buffon, in the fourteenth volume of *Histoire naturelle* (1766), also recognized the obvious physical closeness between humans and apes. But he was not very interested in classification, claiming that there was nothing "natural" about the divisions taxonomists set up. Nature's productions were always "gradual and marked by minute shades."[6] Furthermore, while physically the apes might be intermediate between humans and animals, Buffon maintained that the ape "is in fact, nothing but a real brute, endowed with the external mark of humanity, but deprived of thought, and of every faculty which properly constitutes the human species."[7]

In contrast to these previous views, James Burnett, later known as Lord Monboddo, much to the chagrin of his eighteenth-century Scottish and English Enlightenment colleagues, claimed that some men were born with tails and, furthermore, that the orangutan was a man. In his major treatise *Of the Origin and Progress of Language*, published in 1773, Monboddo maintained that the orang deserved to be included in the family of man for a variety of reasons. Orangs look like us and they also bear the "marks of humanity that I think are incontestable." He alleged such "facts" as "they live in society, they use sticks for weapons, they build huts out of trees, they carry off negroe girls for both pleasure and work, and joined in companies that attack elephants."[8] Although speechless, they had organs for speech. Monboddo's claim met with much criticism. He had not read Tyson's work and thus was quite incorrect when he maintained that the orang had the exact same form as humans and walked upright. While he corrected this oversight in the second edition, he nevertheless continued to argue that the orangutan was "one of us" to support his theory about the origin and development of language.

Fig. 1.
A print of the *HMS Rattlesnake*. Photo reproduced from *T. H. Huxley's Diary of the Voyage of HMS* Rattlesnake edited by Julian Huxley (London: Chatto and Windus, 1935).

Fig. 2.
Thomas Henry Huxley and his wife Nettie, 1882. Photo reproduced from *T. H. Huxley's Diary*, Julian Huxley, editor.

Fig. 3.
French comparative anatomist Baron George Cuvier (1769-1832). Huxley initially found Cuvier's argument in favor of distinct types compelling evidence against transmutation. Wellcome Institute Library, London.

Fig. 4.
German embryologist Karl Ernst von Baer (1782-1876). Huxley adopted von Baer's usage of the type concept in his own work and agreed with von Baer that comparative embryology would provide the key to a natural classification. Ernst Mayr Library of the Museum of Comparative Zoology, Harvard University. © President and Fellows of Harvard College.

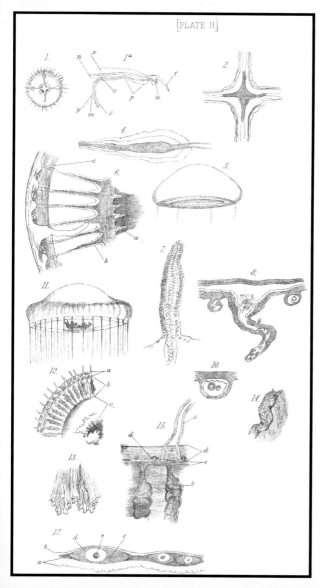

Fig. 5.
One of many of Huxley's own illustrations from "On the Affinities of the Family of the Medusae," 1849, in which he shows that the *Medusae, Monostomatae*, and *Rhizostomidae* are all organized upon a given type. Reproduced from *The Scientific Memoirs of Thomas Henry Huxley*, vol. 1, edited by Michael Foster and E. Ray Lankester (London: Macmillan and Co., 1898).

Fig. 6.
From Huxley's "On the Morphology of the Cephalous Mollusca," 1853. Huxley writes, "The first 11 figures are to be regarded as mere diagrams illustrative of the archetypal form of the Mullusca and its more important modifications. . . . Figs. 9, 10, and 11 are imaginary sections of a Mollusk, a fish, and an articulate animal, respectively, to show the relations of the nervous, alimentary, vascular, and appedicular systems." Reproduced from *The Scientific Memoirs of Thomas Henry Huxley*, vol. 1.

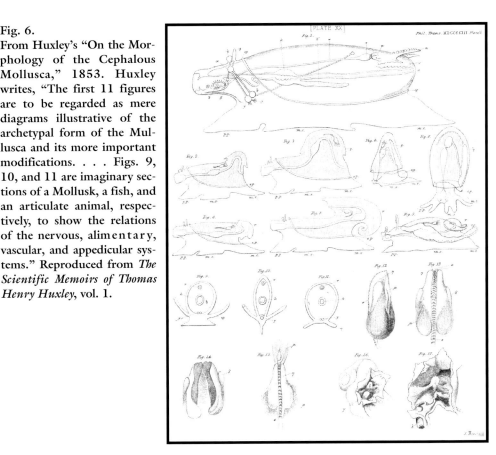

Fig. 7.
A developmental tree from "On the Unity of Structure in the Animal Kingdom," 1837, by Martin Barry. Barry followed von Baer in arguing that development was the key to classification and, like Huxley, believed that form rather than function would provide the deeper insight into solving problems of natural history. Reproduced from *The Meaning of Evolution*, by Robert J. Richards (Chicago: University of Chicago Press, 1992).

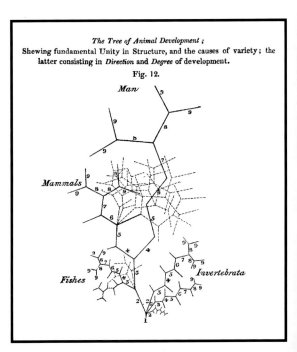

Fig. 8.
Charles Darwin (1809-1882). Huxley defended and disseminated Darwin's theory of evolution, despite his skepticism over various aspects of the theory. © English Heritage Photo Library.

TERTIARY or CÆNOZOIC	Pleistocene	Turbary.		MAN by Remains.		Birds and Mammals.
		Shell-Marl.		by Weapons.		
		Glacial Drift. Bone-Caves.				
		Brick Earth.				
	Pliocene	Norwich ⎫				
		Red ⎬ Crag.				
		Coralline ⎭				
	Miocene	Faluns.		Ruminantia. Quadrumana.	Birds, Orders of.	
		Molasse.		Proboscidia.	Mammals, Orders of.	
	Eocene	Gyps.		Rodentia.		
		London ⎫ Clays.				
		Plastic ⎭		Ungulata. Carnivora.		
SECONDARY or MEZOZOIC	Cretaceous	Maestricht.		Cycloid. ⎫ Fishes. Mosasaurus.		Reptiles.
		Upper Chalk.		Ctenoid. ⎭ Polyptychodon.		
		Lower Chalk.		Birds, by Bones.		
		Upper Greensand.		Procœlian Crocodilia.		
		Lower Greensand.				
	Wealden	Weald Clay.		Iguanodon.		
		Hastings Sand.				
		Purbeck Beds.		Marsupials, — Chelonia by Bones.		
		Kimmeridgian.		Pliosaurus.		
	Oolite U. M. L.	Oxfordian.				
		Kellovian.				
		Forest Marble.				
		Bath-Stone.				
		Stonesfield Slate.		Marsupials.	Amphicoelian Crocodilia. Pterosauria. Homocercal Fishes. Cephalopods 2-gilled.	
		Great Oolite.				
		Lias.		Icthyopterygia.		
	Trias	Bone Bed		MAMMALIA		
		U. New Red Sandstone.		AVES, by Foot-prints.		
		Muschelkalk.		Sauropterygia.		
		Bunter.		Labyrinthodontia.	Crustacea 10-pods.	
PRIMARY or PALEOZOIC	Permian	Marl-Sand.		Sauria.		Fishes.
		Magnesian Limestone.		Chelonia, by Foot-prints.	Isopoda.	
		L. New Red Sandstone.				
	Carbo-niferous	Coal-Measures.		REPTILIA ganoceph. Insecta		
		Mountain Limestone.				
		Carboniferous Slate.				
	Devonian	U. Old Red Sandstone.				
		Caithness Flags.		⎧ ganoid.		
		L. Old Red Sandstone.		⎨ placo-ganoid. Heterocercal.		
		Ludlow		PISCES ⎩ placoid		
	Silurian	Wenlock.				Invertebrates.
		Caradoc.		Echino-derms. Annelids. Bivalves. Trilobites. Pteropods. Brachiopods. Gastropods. Cephalopods 4-gilled.		
		Llandeilo.				
		Lingula Flags.				
		Cambrian.		Fucoids. Zoophytes.		

Fig. 10.
"Table of the Strata and Order of Appearance of Animal Life upon the Earth." Reproduced from *Archetypes and Ancestors*, by Adrian Desmond (Chicago: University of Chicago Press, 1984).

Fig. 9.
British comparative anatomist Richard Owen (1804-1892). Initially on friendly terms, Huxley and Owen were arch-rivals for most of their careers. Wellcome Institute Library, London.

Fig. 11.
Scottish geologist Charles Lyell (1797-1875) wrote the influential *Principles of Geology*. Huxley initially adopted Lyell's antiprogressionist view of the fossil record. Lyell also played a crucial role in Huxley's early saltationist view of evolutionary change. Reproduced from *Sir Charles Lyell's Scientific Journals on the Species Question*, edited by Leonard G. Wilson (New Haven, Conn.: Yale University Press, 1970).

Fig. 12.
Combining both reptilian and birdlike features, *Archeopteryx* was a crucial specimen in Huxley's decision to abandon saltation. Reproduced from *Life's Splendid Drama*, by Peter J. Bowler (Chicago: University of Chicago Press, 1996).

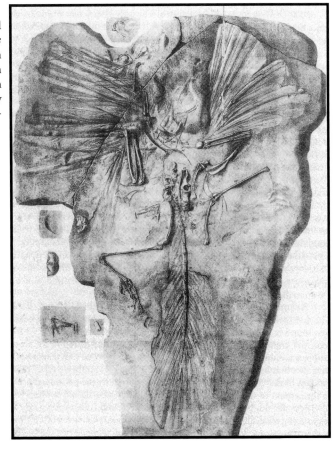

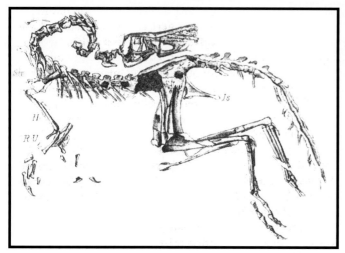

Fig. 13.
The small dinosaur *Compsognathus* with its long, birdlike limbs. Huxley was the first person to suggest that dinosaurs represented the common ancestor and connecting link between birds and reptiles. Reproduced from *Archetypes and Ancestors*, by Adrian Desmond.

Fig. 14.
Thomas Huxley lecturing on the relationship of the great apes to humans. Wellcome Institute Library, London.

Fig. 15.
The anthropomorpha of Karl Linnaeus's *Systema Natural*. The woodcut is taken from the dissertation of his pupil Hoppius and is a mixture of fact and fantasy. Reproduced from *Man's Place in Nature*, by Thomas H. Huxley (New York: D. Appleton, 1898).

Fig. 16.
The "Pygmie," from Edward Tyson's *Orang-Outang, or The Anatomy of a Pygmie Compared with That of a Monkey, an Ape, and a Man*. Although Tyson recognized the creature was a chimpanzee and distinct from both humans and other apes, this figure and that which follows illustrate the confusion surrounding the classification of the "man-like apes" in relationship to humans. Reproduced from *Man's Place in Nature*, by Thomas H. Huxley.

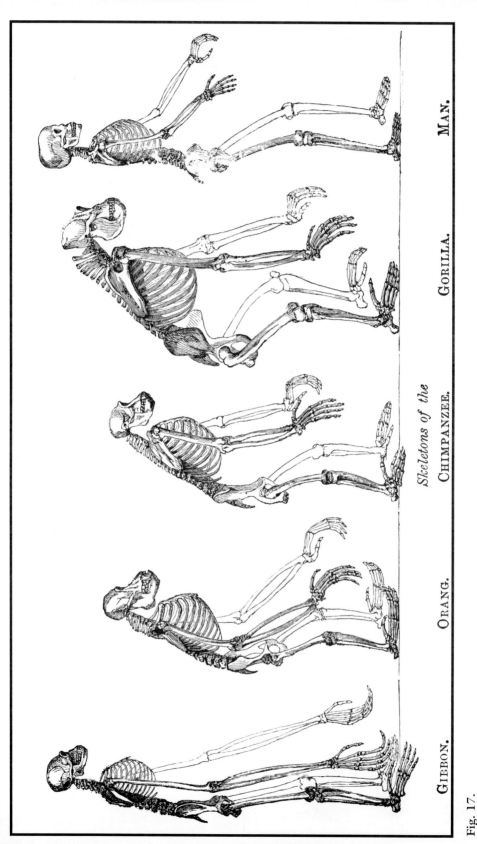

GIBBON.   ORANG.   *Skeletons of the*   GORILLA.   MAN.

CHIMPANZEE.

Fig. 17.
The upright posture of the skeleton emphasizes the closeness of the apes to humans. Reproduced from *Man's Place in Nature*, by Thomas H. Huxley.

The point of this brief and incomplete overview is to demonstrate that there was considerable interest in the "man-like apes" well before the publication of the *Origin*. Their humanlike qualities captured the imagination of naturalists and of the general public alike. Do apes think? Can they develop language? Are they moral beings? With the exception of Lamarck, no one was suggesting the transformation of one species into another. But how to define boundaries between species was actively debated. And certainly the boundary between the manlike apes and humans garnered considerable attention. With Darwin's theory the discussion intensified, not only leading to debates on the origin of *Homo sapiens*, but also rekindling the eighteenth-century controversy over the origin of the different human races: Were human races separate species or just varieties of a single species? With the possibility that humans might actually be related to other primates, the monogenist-polygenist debates became reformulated. Were present-day human races derived from distinct species of ape-like ancestors or just varieties that developed later from one common ancestor?

Huxley certainly reveled in the fray that resulted from the suggestion that humans might be descended from apes. The man-ape question provided an excellent opportunity for him to point out the danger of mixing science and theology. It also brought him into conflict with his enemy Richard Owen. Mario di Gregorio has claimed that the main impetus for Huxley writing *Man's Place in Nature* "was to pursue the denigration of Owen."[9] However, I find his argument unconvincing. Such a view does not do justice to the complexity of Huxley's scientific beliefs or to his relationship with Owen. While the notorious disagreements between Huxley and Owen were real enough, the relationship between the two men had not always been so hostile.

Huxley's Antagonist: Richard Owen

Huxley first came into contact with Owen as a result of difficulties he was having with the government publishing his
research from the voyage of the HMS *Rattlesnake*. Although
two of his papers had been quite favorably received and published while he was still aboard ship, much remained to be
written. Huxley needed time and money to finish his work. He
was told that the Admiralty might be willing to grant him a
nominal appointment on the HMS *Fisguard*, with a leave of
absence from the ship that would allow him to pursue his work
in London. Precedent for such an arrangement existed since
Joseph Hooker had been granted such an appointment to write
up his results from the voyage of the *Erebus* and *Terror*.[10] However, Huxley would need testimonials from leading men of science on the importance of his work for such an appointment to
be even considered. At the time there was no greater man of
science in Britain than Richard Owen. Owen was more than
happy to help the young Huxley and wrote a glowing testimonial to the Admiralty requesting that Huxley have at least a
twelve-month appointment. Huxley had other important men
backing him such as Edward Forbes, Charles Bell, and William
Sharpey, all members of the Committee of Recommendation of
the Royal Society, which allowed him the means to publish his
papers through the Royal Society.[11] Nevertheless, there is no
doubt that Owen's name carried great weight in the Admiralty's decision to give Huxley his appointment.

When Huxley finally met Owen in person, he wrote his
sister Lizzie, "Owen was in my estimation, great from the fact
of his smoking his cigar and singing his song like a brick."[12] Was
Huxley already poking fun at the man he was indebted to? It is
difficult to know for certain. Only a year later he would write
William MacLeay that Owen was "the superior of most, and
does not conceal that he knows it. . . . [He] is an able man, but
to my mind not so great as he thinks himself."[13] Yet Huxley's

letters to his sister and future wife, Nettie, in this same period do not show him as critical of Owen. Rather, they reflect the enthusiasm and idealism of an up-and-coming young man being inducted into the higher echelons of scientific society. He wrote with boyish glee of the famous scientists he was meeting and how proud he felt when they commented favorably on his work. He had found his true calling.

> I have at last tasted what it is to mingle with my fellows—to take my place in that society for which nature has fitted me, and whether the draught has been a poison which has heated my veins or true nectar from the gods, life-giving, I know not, but I can no longer rest where I once could have rested . . . the real pleasure the true sphere, lies in the feeling of self-development in the sense of power and of growing oneness with the great spirit of abstract truth.[14]

Scientific society welcomed the young Huxley. He was elected a Fellow of the Royal Society at the young age of twenty-six, much to his total amazement; he claimed that "they have left behind much better men than I."[15] He thought Owen was probably responsible for the honor and told him so. But Owen replied, "No, you have nothing to thank but the goodness of your own work."[16] Huxley wrote Nettie, "For about ten minutes I felt rather proud of that speech, and shall keep it by me whenever I feel inclined to think myself a fool, and that I have a most mistaken notion of my own capacities."[17]

In the next two years, Owen would write several letters of recommendation for Huxley, and the relationship between the two men was "amazingly civil." However, glimmerings of mistrust and future rivalry existed as well. Huxley described Owen as "a queer fish" and "so frightfully polite," confiding to his sister that he never felt comfortable around the elder man.[18] To William MacLeay he expressed his uneasiness about Owen more explicitly, writing that although Owen had helped him, he was a man "with whom I feel it necessary to be always on my guard."[19] Huxley was not alone in his distrust of Owen and was astonished at the degree of animosity toward the man, claiming that he was both "feared and hated." How was Huxley to deal with a person who was very powerful, on whom he was dependent for testimonials, and yet someone he did not trust?

Huxley's relationship with Owen was further complicated by the state of British science at the time. Huxley had been an outstanding success. Not only had he been elected to the Royal Society, but he was awarded its medal the following year. In spite of this, he was still haggling over who was going to pay for the publication of his research, and he did not have a permanent job. He was penniless and wanted to marry, but it appeared that a life of science might be incompatible with earning a living. At the time, probably only four or five positions existed in London for zoologists or comparative anatomists, and the pay was equivalent to that of a civil servant. Owen, whose European reputation was second only to Cuvier's, received only £300 as Hunterian Professor.[20] Edward Forbes, another leading paleontologist, received the same salary. With so few positions available, competition, jealousy, and underhanded dealings were rife in the scientific community, and Huxley's idealism was soon squashed. "Science is I fear, no purer than any other region of human activity; though it should be. Merit alone is very little good; it must be backed by tact and knowledge of the world to do very much."[21]

Although intrigues and rivalries existed throughout the British scientific community, in large measure they appeared to center around one man: Richard Owen. At the time, the scientific world was astir with who was to take over Charles König's place as keeper of mineralogy and geology in the British Museum and whether a separate institution would be established just for zoology, perhaps with Owen as its head. Huxley found the heart-burnings and jealousies about the matter beyond conception. He wrote MacLeay that if Owen and John

Edward Gray, who was the keeper of zoology, were to work in the same institution, "it [was] predicted that in a year or two the total result will be a caudal vertebra of each remaining after the manner of the Kilkenny cats."²² Huxley claimed that the majority of Owen's contemporaries intensely hated Owen and admitted the man did "some very ill-natured tricks now and then." The latest example was Owen's article criticizing Lyell's 1851 anniversary address as president of the Geological Society. It was not just that he attacked Lyell's progression, but he also ridiculed J. T. Quekett, whom Huxley described as an inoffensive man and Owen's subordinate. The review was "not more remarkable for its severity than for its bad taste" and had harmed Owen's reputation considerably.²³ Huxley was having his own problems with Owen as well. The Royal Society had not provided funds for Huxley's publications, claiming that since he was still in government service, it was the duty of the Admiralty to provide funds. Therefore, in November 1852, Huxley asked Owen for another testimonial to send directly to the Secretary of State to try and secure funding. After ten days he still hadn't heard anything. He complained to Forbes, who told him to ask Owen again. He wrote Owen and again heard nothing. Huxley was furious and was going to walk right past Owen several days later when he ran into him in the street.

But he stopped me, and in the blandest and most gracious manner said "I have received your note. I shall *grant* it." The phrase and the implied condescension were quite "touching"—so much that if I stopped a moment longer I must knock him into the gutter. I therefore bowed and walked off. This was last Saturday. Nothing came on Monday or on Tuesday, but on Wednesday morning I received "with Prof. Owen's best wishes," *the strongest and kindest testimonial any man could possibly wish for*! I could not have dictated a better one.²⁴

What was Huxley to make of Owen's contradictory behavior? The problem was further complicated because although Huxley still needed Owen for testimonials, at the same time, on his own subjects Huxley already considered himself Owen's master. And for all of Huxley's claims of disgust as to the scheming and strategies involved in becoming a success, he took "a certain pleasure in overcoming these obstacles and fighting these folks with their own weapons."[25] Huxley was being initiated into the politics of science and was more than willing to play the game to ensure his own success. As for Owen, he claimed, "I am quite ready to fight half a dozen dragons. And although he has a bitter pen, I flatter myself that on occasions I can match him in that department also."[26] Huxley's words were prophetic.

In 1854, Huxley finally succeeded in securing a permanent position in London. Edward Forbes left his position at the London School of Mines for a position in Edinburgh. Earlier, he had asked if Huxley would be willing to replace him if he left. Huxley took over one of his lectureships and was also appointed to be in charge of a coast survey for the Geological Survey. Shortly afterward he was offered a lectureship on comparative anatomy at St. Thomas's Hospital. In addition, several mono-graphs on his research from the *Rattlesnake* were published.

Huxley was becoming more and more successful, and it appears that Owen was quite threatened by the younger man. The final strain in their association occurred in 1857. Owen was giving a paleontology course at the lecture theater of the School of Mines and was listed in the medical directory as Professor of Paleontology at the School of Mines. Huxley wrote John Churchill of the directory, pointing out that "Mr. Owen holds no appointment whatsoever at the Government School of Mines." Since Huxley was the Professor of General Natural History, which included comparative anatomy and paleon-tology, he asked why Owen was listed with such a title.[27] Churchill's explanation was totally unsatisfactory, indicating that "the designations attached to Mr. Owen's name in the

Medical Directory are inserted correctly from the return as furnished by that gentleman."[28] Huxley was furious, perceiving Owen's actions as a direct threat to his own position at the School of Mines, and broke off his relationship with him.[29]

The obvious personal antagonism between the two men, however, should not obscure the real scientific differences which existed between them from the beginning of their relationship. Fundamentally, Huxley disagreed with the philosophical underpinnings of Owen's science. Owen was firmly in the tradition of the idealistic German morphologists. Although Huxley had also been trained in this tradition, he quickly distanced himself from it, as has been pointed out more thoroughly in chapter 2. Huxley had absolutely no use for the wider and sometimes mystical speculations of the transcendental anatomists. In "On the Cephalous Mollusca," Huxley pointed out that his use of the word "archetype" was quite distinct from that of the transcendentalists, and it was apparent that his comments were directed against Owen's use of the word.[30] It is in light of this fundamental disagreement that Huxley's 1858 Croonian Lecture "On the Theory of the Vertebrate Skull" should be examined. Describing Lorenz Oken as a "fanciful philosopher," Huxley demolished his view promulgated by Owen that the skull was a modified vertebrae. Based on the work of Martin Rathke, Karl Reichert, and others, Huxley presented embryological evidence demonstrating that the development of the skull and the vertebrae were quite different. "The spinal column and the skull start from the same primitive identity . . . whence they immediately begin to diverge. . . . It is no more true that the adult skull is a modified vertebral column than it would be to affirm that the vertebral column is a modified skull."[31] Huxley undoubtedly took great pleasure in discrediting the views of a man he intensely disliked. But this should not overshadow the scientific disputes between the two men or Huxley's deep interest in the problem of development. As Sir Michael Foster later wrote,

> This lecture marked an epoch in England in vertebrate morphology and the views enunciated in it carried forward, if somewhat modified . . . not only by Huxley's subsequent researches and by those of his disciples, but especially by the splendid work of [Carl] Gegenbaur, are still in the main the views of anatomists of today.[32]

Owen's science was outdated; his theological beliefs caused him to espouse views that were becoming increasingly untenable in light of modern research. His religious beliefs regarding the uniqueness of humans combined with his idealist morphology led him to propose that man should be placed in a separate division of his own and resulted in his most famous conflict with Huxley—over the presence of a hippocampus in the brains of nonhuman primates.

The Hippocampus Debacle

Taxonomists intensely debated how the human species should be classified. What was the relationship of humans to other primates? Most people were deeply opposed to anything that broke down the barrier between humans and the rest of the animal world. Huxley claimed that his own mind was not made up in 1857 when Owen gave a paper to the Linnaean Society entitled "On the Characters, Principles of Division, and Primary Groups of the Class Mammalia."[33] Owen asserted that certain anatomical features of the brain were unique to the genus *Homo*, and therefore he wanted to place him in a new division, "Archencephala," that was a separate genus superior to the others in the mammalian class. Huxley disagreed, based on the work of earlier anatomists, and set about to investigate the matter for himself.[34] He found that the earlier researchers were correct and that humans had no unique structures from the higher apes, and many features were shared by the lower apes as well. Although

Huxley did not make any public comment on his views at this time, he incorporated his findings in his teachings and continued to research the matter over the next two years.

Thus, Huxley was already investigating the taxonomy of apes and humans before the publication of the *Origin*. Michael Bartholomew has claimed that Huxley's work underwent no fundamental shift when the *Origin* appeared,[35] but this was because the *Origin* strongly supported Huxley's findings regarding the structural relationships between humans and apes by providing a causal explanation as to why these similarities existed. Not only was a shift in thinking unnecessary to incorporate the ideas embodied in the *Origin*, but Darwin's treatise inspired him to continue his investigation. In the preface to the 1894 edition of *Man's Place in Nature*, Huxley wrote, "And in as much as development and vertebrate anatomy were not among Mr. Darwin's many specialties, it appeared to me that I should not be intruding on the ground he had made his own, if I discussed this part of the general question. In fact, I thought that I might probably serve the cause of evolution by doing so."[36] One might suspect that Huxley was engaging in a bit of revisionist history here, but there are good reasons to accept Huxley's assertion at face value.

Huxley received further stimulus to continue his investigations as a result of his famous encounter with Bishop Samuel Wilberforce at Oxford in 1860, in which the bishop asked Huxley whether "it was on his grandfather's or his grandmother's side that the ape ancestry comes in."[37] Huxley replied

that a man has no reason to be ashamed of having an ape for his grandfather. If there were an ancestor whom I should feel shame in recalling it would rather be a man—a man of restless and versatile intellect—who, not content with an equivocal success in his own sphere of activity, plunges into scientific questions with which he has no real acquaintance, only to obscure them by an aimless rhetoric, and distract the attention of his hearers from the real point at issue by eloquent digressions and skilled appeals to religious prejudice.[38]

"Soapy Sam" Wilberforce had met his match in the young Huxley. Both Huxley and Darwin's theory would have to be taken seriously. The event was a significant point in Huxley's career. It was also an extremely important event for Darwin's theory. As Leonard Huxley wrote,

> The importance of the Oxford meeting lay in the open resistance that was made to authority, at a moment when even a drawn battle was hardly less effectual than acknowledged victory. Instead of being crushed under ridicule, the new theories secured a hearing, all the wider, indeed, for the startling nature of their defense.[39]

While recent scholarship has correctly criticized the overly simplistic analysis of the conflict between science and religion in such books as John Draper's *History of the Conflict Between Religion and Science* and Andrew White's *A History of the Warfare of Science with Theology*, one should not underestimate the importance of this conflict. Draper was at the Oxford meeting, and it was not a coincidence that he wrote his book in 1875, when many Victorians were still in an uproar over Darwinism's implications for Christianity. The role of God, the question of "man's place in nature," was, in fact, precisely why the *Origin* was so controversial. Many of the scientific objections brought forth by people such as George Mivart, Louis Agassiz, and the Duke of Argyll boiled down, in the last analysis, to the question "Where was God in the *Origin*?" Huxley was attracted to the *Origin* precisely because it removed investigations about the history of life from the realm of theology. This was a far more powerful reason for him to become embroiled in the controversy over human origins than merely to engage in a personal vendetta against Owen. Indeed, it was Owen's mixing of science with theology that led him into a completely untenable position regarding the structure of the human brain. Huxley could show the dangers of mixing theology with science, he could defend evolution, and he could discredit Owen, all by investigating the ape-human question—an opportunity far too good to pass up.

Huxley presented his findings on the relationship of humans to other primates in "On the Zoological Relations of Man with the Lower Animals."[40] This paper embodies the strategy that Huxley used again and again in his support of evolutionary theory. He maintained that the question of classification was a purely scientific one. He did not want questions of the origins of man's morals or ethics mixed in with questions of anatomy. Huxley believed that an unequivocal demonstration of the close relationship between apes and humans would be the most powerful support for Darwin's theory, because Darwin provided a clear, logical explanation for the existence of those relationships. Huxley acknowledged that the question of human classification was controversial. Linnaeus had classified humans and apes in the same order, primates, with the genera *Homo*, *Sima* (also *Simia*), *Lemur*, and *Vespertilio*—all having equal ranking. Cuvier believed that because humans had two hands instead of four, as well as other distinguishing characteristics, this justified classifying them as a distinct order, which he named *Bi-mana*. Owen wanted *Homo* to be classified as a distinct subclass, and Terres thought humans were so distinctive that they should have their own kingdom equal to Animalia or Plantae. Some people even thought humans should not be thought of zoologically at all.[41] "Ingenious and learned men" held such a wide range of opinion that Huxley believed the problem was "in reality more one of opinion as to the right method of classification and the value of groups which receive certain names than one of fact."[42] The classification of *Homo sapiens* had been complicated because "passion and prejudice have conferred upon the battle far more importance than, as it seems to me, can rationally attach to its issue."[43] For Huxley, the question should be strictly a scientific one that could be resolved by the facts of comparative anatomy and physiology, "independently of all theoretical views."[44] Darwin's theory, special creation, or any other theory need not even be mentioned in the investigation of the "facts" of anatomy and physiology. After such facts were established, then one could comment on how well a particular theory fit the facts. Huxley, moreover, believed the facts were well known. The researches of Duvernoy, Fredrich Tiedemann, Isiodor St. Hilaire, Jacobus Schroeder Van der Kolk, Willem Vrolik, Louis Gratiolet, and Owen all pointed to the same conclusion—that apes and humans were extremely closely related. But Huxley realized that before he could discuss anatomical and physiological evidence, he had to convince his audience that such a view in no way detracted from the dignity of humankind. It is this aspect of Huxley's writings that is so powerful, setting his work apart from other works on the ape-human question.

Huxley claimed that it didn't matter whether man's origin was distinct from all other animals or whether he was the result of modification from another mammal: "His duties and his aspirations must, I apprehend, remain the same. The proof of his claim to independent parentage will not change the brutish-ness of man's lower nature nor . . . will man's pithecoid pedi-gree one whit diminish man's divine right of kinship over nature."[45] Human dignity, according to Huxley, was not inher-ited, but rather "to be won by each of us so far as he con-sciously seeks good and avoids evil, and puts the faculties with which he is endowed to its fittest use."[46] Thus, for Huxley, all aspects of human nature, both "brutishness" and "princely dig-nity," would have to be accounted for independently of the question of human origins. If he could convince his readers that the highly charged issues concerning man's morals and ethics, questions of good and evil, were not relevant to the question of human origins, the problem of classification could be investi-gated objectively and dispassionately.

This is not to say that Huxley was uninterested in such questions. In fact, in his 1893 Romanes lecture on evolution and ethics, Huxley has provided us with one of the clearest articulations of the problem of evolutionary ethics. Even if one accepts that evolution has produced creatures such as ourselves with a moral sense, it does not follow that we can look to evolution to define the content of what we call moral.

> The propounders of what are called the "ethics of evolution," when the "evolution of ethics" would usually better express the object of their speculations, adduce a number of more or less interesting facts and more or less sound arguments in favor of the origin of the moral sentiments, in the same way as other natural phenomena, by a process of evolution. . . . But as the immoral sentiments have no less been evolved, there is so far, as much natural sanction for the one as the other. The thief and the murderer follow nature just as much as the philanthropist. Cosmic evolution may teach us how the good and the evil tendencies of man may have come about; but, in itself, it is incompetent to furnish any better reason why what we call good is preferable to what we call evil than we had before.[47]

I will return to the Romanes lecture at the end of the chapter. For now, however, let us examine Huxley's strategy for convincing people that humans were no exception to Darwin's theory.

Huxley did not directly address the problem of whether humans were descended from an apelike ancestor. Instead, he asked how closely related apes and humans were, approaching the question just as a taxonomist would investigate how closely related were the cat and the dog. It is clear, however, from the rhetorical structure of the lectures that the close taxonomic relationship between humans and other primates was intended to be interpreted as support for human evolution from lower animals.

Like Darwin, Huxley appealed to the Victorians' love of their pets to demonstrate the unity of humans with the animal world.

The dog, the cat . . . return love for our love and hatred for
our hatred. They are capable of shame and sorrow; and
though they may have no logic nor conscious ratiocination,
no one who has watched their ways can doubt that they possess that power of rational cerebration which evolves reasonable acts from the premises furnished by the senses.[48]

Thus, Huxley claimed a psychical as well as physical unity
existed between man and beast. And who did Huxley cite in
support of such a view? None other than his arch antagonist,
whom he directly attacks only a few pages later—Richard
Owen. In one of his more racist statements, Owen had written,

Not being able to appreciate or conceive of the distinction
between the psychical phenomena of a chimpanzee and a
Boschisman or of an Aztec, with arrested brain-growth, as
being a nature so essential as to preclude a comparison between them, or as being other than a difference of degree, I
cannot shut my eyes to the significance of that all-pervading
similitude of structure—every tooth, every bone, strictly
homologous—which makes the determination of the difference between *Homo* and *Pithecus* the anatomist's difficulty.[49]

Owen removed this note in a later reprint of the paper, undoubtedly realizing it provided support for a position in direct
opposition to the views he held. He believed that man's "psychological powers, in association with his extraordinarily developed brain, entitle the group which he represents to equivalent
rank with the other primary divisions of the class *Mammalia*,
founded on cerebral characters."[50] Huxley then proceeded to
demolish Owen's view that the human brain exhibited peculiar
and unique characters.

Huxley first discussed the embryological development of
humans and the higher apes as well as presented a detailed comparison of the hand, foot, and brain. All the evidence showed
that the differences between humans and the higher apes were

no greater than those between the higher and lower apes. Citing the research of Tiedemann, St. Hilaire, and as previously mentioned even Owen himself, he claimed that the differences between humans and apes were no greater than in the other genera of the Quadrumana. Owen had gotten himself into a totally untenable position by claiming that *Homo sapiens* had a unique structure, the hippocampus minor, which justified ranking it in its own subclass. Unlike Huxley, he needed to ground his belief in the uniqueness of the human psyche in human anatomy. Huxley showed no mercy in his attack on Owen. He quoted Owen extensively, only to discredit him.

> In man the brain presents an ascensive step in development.
> . . . Not only do the cerebral hemispheres overlap the olfac-
> tory lobes and cerebellum, but they extend in advance of the
> one and further back than the other. Their posterior develop-
> ment is so marked that anatomists have assigned to that part
> the character of a third lobe; it is peculiar to the genus *Homo*
> and equally peculiar is the posterior horn of the lateral ven-
> tricle and the hippocampus minor which characterized the
> hind lobe of each hemisphere.[51]

But Huxley then demonstrated in great detail that none of Owen's statements was true. The third lobe and the posterior cornu of the lateral ventricle were "neither peculiar to nor characteristic of man." Both traits were found in all of the higher Quadrumana. The hippocampus was also not unique to humans, having been found in certain higher Quadrumana.[52] Huxley cited the work of Schroeder van der Kolk, Willem Vrolik, and others in support of his views. He quoted a recent letter from Allen Thomson summarizing the result of his dissection of the brain of a chimpanzee. Thomson stated:

> There is very clearly, a posterior lobe, separated from the
> middle one by as deep a groove between the convolutions of
> the inner side of the hemispheres as in man, and equally well

marked off on the outer side. I should be inclined to say that the posterior lobe is little inferior to that of man. . . . I found an eminence in the floor of the posterior cornu and towards its inner side which I regard as the hippocampus minor and I found it produced in exactly the same manner as in man.[53]

Huxley concluded that "every original authority testifies that the presence of a third lobe in the cerebral hemisphere is not 'peculiar to the genus *homo*,' but that the same structure is discoverable in all the true *Simiae* . . . and is even observable in some lower mammalia."[54]

However, Owen would not give up. On March 19, 1861, he gave a lecture at the Royal Institution in which he still maintained that only humans had a hippocampus minor and that the difference between the human brain and a gorilla or chimpanzee was far greater than between the brains of other primates. The lecture was reported in the *Athenaeum*, and Huxley responded with a letter pointing out the many errors in Owen's views, including commenting on the incompleteness and inaccuracy of the diagram of a gorilla brain that appeared in the article. A series of letters were exchanged in which Owen seemed determined to make a fool of himself. Huxley described the inanity of the dispute to Hooker:

> A controversy between Owen and myself which I can only call absurd (as there is no doubt whatever about the facts) has been going on in the *Athenaeum* and I wound it up in disgust last week. . . . I do not believe that in the whole history of science there is a case of any man of reputation getting himself into such a contemptible position. He will be the laughing stock of all the continental anatomists.[55]

The matter was finally put to rest in 1862 when Sir William Flower in a public dissection of ape brains demonstrated the existence of those cerebral characteristics that were claimed to be unique to man.

Nicolaas Rupke has argued that Huxley's victory was not as complete as Huxley and the pro-Darwin forces would have us believe. He points out that Owen had never denied that the gorilla had a hippocampus minor, but that it was not as well developed as in humans and so did not deserve the name "hippocampus minor." But Rupke also admits that by claiming the "posterior lobe," "posterior cornu," and "hippocampus minor" were features that were characteristic of the human brain, Owen implied that these features were not present in simian brains. Furthermore, Owen had not made clear that "indications" or "traces" of a posterior cornu and hippocampus minor were present in apes or monkeys in his original 1857 lecture. Owen later retreated to a more moderate position, but Huxley's response was to this 1857 lecture. Rupke also confirms that new anatomic information, particularly Flower's study, showed that the cerebral characteristics did not vary in the distinct manner that Owen had argued and thus were not very useful for taxonomic purposes. Therefore, while Rupke's account of this controversy is more sympathetic to Owen than my own, it confirms my own analysis.[56]

Man's Place in Nature

In spite of the heated debate between the two men, it is important to emphasize that Huxley was not just carrying on a dispute with Owen. The circumstances that led to the writing of *Man's Place in Nature* illustrate that scientific disputes, personality conflicts, and differences in worldview were inextricably intertwined in the question of human ancestry. As a result of the publication of the *Origin* and his clash with Wilberforce, in the spring of 1861 Huxley decided to devote his weekly lectures to working men to "The Relation of Men to the Rest of the Animal Kingdom." He was committed to educating people about Darwin's theory, and he knew that the most controversial aspect of it was the question of human ancestry. The lectures were quite popular. He wrote his wife, "Lyell came and was rather astonished at the magnitude and attentiveness of the audience," and he quipped that "by next Friday evening they will all be convinced that they are monkeys."[57] In the midst of all the controversy over the hippocampus, much to Huxley's surprise he was invited to give two lectures on "The Relation of Man to the Lower Animals" for the Philosophical Institute of Edinburgh. Delivered on January 4 and 7, 1862, these lectures were a less technical presentation of his earlier paper. He explicitly claimed that the comparative anatomy of primates provided powerful evidence in favor of Darwin's theory. The gaps between humans and the higher apes were no greater than between the higher apes and the lower apes. If Darwin's hypothesis explained the common ancestry of the latter, then it followed that it also explained the origin of the former.[58] The lectures were a success, but Huxley felt they had not reported quite accurately his comments about Darwin's theory. He wrote Darwin: "Nor have they reported here my distinct statement that I believe man and the apes to have come from one stock."[59] To Hooker he complained of the same oversight: "I told them in so many words that I entertained no doubt of the origin of man from the same stock as the apes. . . . The report does not put nearly strong enough what I said in favour of Darwin's views. I affirmed it to be the only scientific hypothesis of the origin of species in existence."[60]

Nevertheless, Huxley was pleased with the reception he received in Edinburgh. These lectures allowed him to promulgate Darwinism, and in doing so, he was once again able to reiterate the importance of keeping scientific questions distinct from theological ones. That he was even invited to "saintly" Edinburgh and could speak to an audience who applauded his belief that humans had an ape-like ancestor was a "grand indication of the general disintegration of old prejudices which is going on."[61] Huxley hoped to develop the lectures further, give them again in London, and then publish them. They did eventually form part 2 of *Man's Place in Nature*.

Huxley published *Man's Place in Nature* in 1863 in spite of warnings that it would be the undoing of his career. The book consisted of three parts. Part 1, "On the Natural History of the Man-Like Apes," provided an overview of how the great apes had been regarded through history. It illustrated that the problem of the relationship of apes to humans had fascinated people for centuries. Do apes think? Are they moral beings? Do they have societies? Facts and mythology had intermingled, resulting in much confusion over the classification of these animals. Although not directly stated, Huxley was implying that an objective investigation of these animals, carefully examining their physical makeup, would clear away much of the confusion, thus laying the groundwork for part 2 of his book.

The heart of *Man's Place in Nature*, both literally and figuratively, was part 2 of the book. Huxley had refined his earlier lectures "On the Relation of Man to the Lower Animals," and the result was an educational tour de force documenting "the extent of the bonds which connect man with the brute world."[62] He then used this evidence to argue for the superiority of the Darwinian hypothesis to all others regarding the origin of humans. Huxley's presentation of the facts of development and comparative anatomy was superb. It was hard to imagine how anyone could deny their ape ancestry after reading his discussion. But Huxley knew beliefs run deep. Scientists or members of the working class, devout or not, most Victorians would still be appalled at the inevitable conclusions to be drawn from Huxley's analysis. They would argue that "the belief in the unity of origin of man and brutes involves the brutalization and degradation of the former."[63] But Huxley questioned if that were really so. In a passionate entreaty, he claimed that man's dignity did not depend on his physical characteristics or his origins:

> It is not I who seek to base Man's dignity upon his great toe or insinuate that we are lost if an Ape has a hippocampus minor. . . . On the contrary, I have done my best to sweep away this

vanity. . . . Is it, indeed, true, that the Poet, or the Philosopher, or the Artist whose genius is the glory of his age, is degraded from his high estate by the undoubted historical probability, not to say certainty, that he is the direct descendant of some naked and bestial savage, whose intelligence was just sufficient to make him a little more cunning than the Fox, and by so much more dangerous than the Tiger? Or is he bound to howl and grovel on all fours because of the wholly unquestionable fact, that he was once an egg, which no ordinary power of discrimination could distinguish from that of a Dog? Or is the philanthropist, or the saint, to give up his endeavours to lead a noble life, because the simplest study of man's nature reveals, at its foundations, all the selfish passions, and fierce appetites of the merest quadruped? Is mother-love vile because a hen shows it, or fidelity base because dogs possess it?[64]

In this passage, Huxley was obviously attacking Owen, who claimed that not only the hippocampus minor, but also the hallux, or big toe, was unique to *Homo sapiens*. However, it would be a mistake to regard this as just another example of Huxley indulging in Owen-bashing. His point was substantive: Just because we share admirable traits with the lower animals does not make them less admirable. Furthermore, for Huxley, man's lowly ancestry was "the best evidence of the splendour of his capacities."[65] He ended part 2 by drawing an analogy between civilized man and the magnificent alpine mountains. Like the mountains, which were the hardened mud of primeval seas that had risen to their glorious state, humans had also risen from their lowly state. Huxley implored his readers to put passion and prejudice aside and recognize that

our reverence for the nobility of manhood will not be lessened by the knowledge that Man is, in substance and in structure, one with the brutes; for, he alone possesses the marvellous endowment of intelligible and rational speech, slowly accumulated and organized the experience which is almost wholly lost with the cessation of every individual life in other animals; so that, now, he stands raised upon it as on a mountain top, far above the level of his humble fellows, and transfigured from his grosser nature by reflecting here and there, a ray from the infinite source of truth.[66]

Huxley may have thought we descended from the brutes, but it is clear he also believed that a vast gap separated us from the rest of the animal world. Man's pithecoid ancestry did not detract from his noble possibilities. For Huxley, it made him an even more remarkable creature.

Part 3 of *Man's Place* discussed the recent discoveries of the fossil remains of man, including the Engis skull discovered by Schmeding in the Valley of the Meuse in Belgium, and Neanderthal man. Huxley concluded from these ancient skulls that they were not significantly different from modern humans and thus "do not take us appreciably nearer to that lower pithecoid form, by the modification of which he has probably become what he is."[67] He asked his readers, "Where then, must we look for primaeval Man: Was the older *Homo sapiens* pliocene or miocene, or yet more ancient?"[68] Only time would tell. But Huxley believed that the finds of paleontology would push the origin of humans back to a far earlier epoch than anyone had previously imagined. If humankind was that ancient, clearly this was evidence against the creation hypothesis. Once again the Darwinian hypothesis was the only proposition that could make sense of these ancient human fossils.

The Impact of Man's Place in Nature

What was the impact of *Man's Place in Nature* in Huxley's own time? In his attempt to validate his assertion that the book was written primarily to discredit Owen, di Gregorio claims that it only became well known much later and was recognized as an important "forward looking treatise."[69] The *Quarterly Review* did not bother to review it, and "the critical reviews display much more fairness and much less excitement than those of *The Origin of Species* and the later *Descent of Man*."[70] However, Leonard Huxley pointed out, as did di Gregorio in an earlier part of his book, that *Man's Place in Nature* was an immediate success. Published in January 1863, by the middle of February it had sold two thousand copies. By July it was republished in America and was being translated into French and German.[71] It was steadily reprinted for the next forty years. To leading men of science, the book had made an important impression. Darwin wrote Lyell, "How splendid some pages are in Huxley."[72] Lyell replied, "If he had leisure like you and me;—and the vigour and logic of the lectures; and his address to the Geological Society, and half a dozen other recent works (letters to the 'Times' on Darwin, &c.), been all in one great book, what a position he would occupy!"[73] Clearly, Huxley's book was not going unnoticed.

An alternative and no less important assessment of the impact of *Man's Place in Nature* is to examine the popular literature surrounding the debate over human ancestry. The discussions over evolution and human origins had spread far beyond the staid halls of the Royal and Geological Societies. The hippocampus debate represented more than some esoteric scientific disagreement over an obscure part of the brain. Rather, "man's place in nature" was at stake. What made humans unique? What was the role of science in answering such a question? How could evolutionary ideas be made compatible with theology? It is certainly true that the personality clash between Owen and Huxley played a significant role in the popular accounts. London society reveled in the squabbles between the two men, and satirical accounts soon appeared about the great hippocampus debate. However, the fact that their disputes were the subject of popular satire indicates how deeply the ape-human question had permeated society.

Punch printed a poem supposedly authored by a gorilla

from the zoological gardens entitled "Monkeyana" with a picture of an apelike creature carrying a sign asking "Am I a Man and a Brother?" The poem was about Darwin's theory, but half of it spoofed the Owen-Huxley clash:

> Then HUXLEY and OWEN,
> With rivalry glowing,
> With pen and ink rush to scratch;
> 'Tis Brain *versus* Brain
> Till one of them's slain;
> By Jove! it will be a good match!
>
> Says OWEN you can see
> The brain of Chimpanzee
> Is always exceedingly small,
> With hindermost "horn"
> Of extremity shorn;
> And no "Hippocampus" at all.
>
> The Professor then tells 'em,
> That man's "cerebellum,"
> From a vertical point you can't see;
> That each "convolution"
> Contains a solution,
> Of "Archencephalic" degree
>
> Then apes have no nose.
> And thumbs for great toes,
> And a pelvis both narrow and slight;
> They can't stand upright,
> Unless to show fight,
> With "Du Chaillu,"* that chivalrous knight!

---

*Paul du Chaillu was an explorer and anthropologist who provided gorillas and gorilla parts to the museum world of London and who became a protégé of Owen. In his book *Exploration and Adventure* du Chaillu described a confrontation with a large male gorilla, but soon people were taking sides as to the accuracy of the account. The controversy surrounding du Chaillu became linked to the political battles between Owen and John Gray over the establishment of a separate British museum for natural history as well as the hippocampus episode between Owen and Huxley. For a full account of the affair see Rupke, *Victorian Naturalist*, pp. 215–22.

Next HUXLEY replies,
That OWEN he lies,
And garbles his Latin quotation;
That his facts are not new,
His mistakes not a few,
Detrimental to his reputation,

"To twice slay the slain,"
By dint of the Brain,
(Thus HUXLEY concludes his review)
Is but labour in vain,
Unproductive of gain,
And so I shall bid you "Adieu!"[74]

This poem reveals that the debates over human ancestry also had implications for the relationship between the races. The United States was immersed in a civil war over the issue of slavery, with the antislave forces adopting "Am I a Man and a Brother?" as their slogan. Are not Negroes our brothers? Certainly if a gorilla is claiming to be related to humans, what does this say about all of humankind?

In 1863, an anonymously printed pamphlet called *A Sad Case* (it was, in fact, published by George Pycroft) appeared, describing the fictitious courtroom proceedings of Owen and Huxley, who had been charged with disturbing the peace. The two men had been about to come to blows when the police arrived. Huxley had called Owen a "lying Orthognathous Brachcephalic Bimaneous Pithecus," while Owen had accused Huxley as being "nothing else but a thorough Archencephalic Primate."[75] As the trial proceeded, it was apparent that the dispute was not just about whether apes had a hippocampus minor. Owen was outraged that he was being told in public that he was physically, morally, and intellectually only a little better than a gorilla.[76] However, the problem went even deeper than that. Owen claimed that he had regarded Huxley as a "quiet well-meaning man," but since he had become successful he had become "highly dangerous."⁷⁷ Furthermore, Huxley and his new pals, Darwin, Rolleston, and others were always ridiculing him. But Huxley countered that everything was fine "as long as Dick Owen was top sawyer, and could keep over my head, and throw his dust down in my eyes. There was only two or three in our trade, and it was not very profitable; but that was no reason why I should be called a liar by an improved gorilla, like that fellow."⁷⁸ A Mr. Bull testified that there was bad feeling between the whole lot of these scientists. "Huxley quarreled with Owen, Owen with Darwin, Lyell with Owen, Falconer and Prestwich with Lyell, and Gray, the menagerie man, with everybody."⁷⁹ The mayor suggested that perhaps the clergy might exert some influence over them, to which Mr. Bull replied that "no class of men paid so little attention to the opinions of the clergy as that to which these unhappy men belonged."⁸⁰ The story ended with the mayor admonishing both men and telling them that they should be friends and help each other in their work. But as soon as they were out the door, an altercation ensued. That night Huxley was again attacking Owen in his lecture on man's place in nature. "But as the assemblage consisted of working men, and as they were very orderly they were not interfered with by the police."⁸¹

A Sad Case wonderfully illustrates the various issues surrounding the controversy over human origins. While the focal point was the clash between Huxley and Owen, their disagreements reflected the fighting for power and prestige that was occurring within the scientific community as a whole. Were Darwinian ideas going to prevail? Who would obtain the few paying jobs that were available? Was science responsible for the loss of faith? Huxley promulgated his ideas to the common man by means of his working men's lectures. Everything was grist for the mill in this astute satire.

The most famous account of the feud between Huxley and Owen is undoubtedly that written by Charles Kingsley, who immortalized the dispute in his satirical fairy tale *The Waterbabies*, sparing neither man in his commentary. In the first part of the book Huxley and Owen were mentioned as experts who could help evaluate whether waterbabies really existed. Surely if waterbabies existed, one would have been caught, put in spirits (formaldehyde) and cut in half. One half would have been given to Professor Huxley and the other to Professor Owen to see what each had to say about it.[82] Later in the book, the two men appeared as caricatures of themselves. The empiricist Huxley was portrayed as the great naturalist Professor Pttmilnsprts of Necrobioneopaleonthydrochthoanthropopithekology, member of the Acclimatisation Society. He held the view that "no man was forced to believe anything to be true, but what he could see, hear, taste, or handle." Furthermore, "he had even got up once at the British Association and declared that apes had hippopotamus majors in their brains just as men have."[83] But clearly, Kingsley was on Huxley's side. Ridiculing Owen's view, the narrator continued

> [that] was a shocking thing to say; for if it were so what would become of the faith, hope, and charity of immortal millions? You may think that there are other more important differences between you and an ape, such as being able to speak and make machines, and know right from wrong, and say your prayers, and other little matters of that kind: but that is a child's fancy my dear. Nothing is to be depended on but the great hippopotamus test. If you have a hippopotamus major in your brain, you are no ape, though you had four hands, no feet, and were more apish than the apes of all aperies. But if a hippopotamus major is ever discovered in one single ape's brain, nothing will save your great-great-great-great-great-great-great-great-great-great-greater-greatest grandmother from having been an ape too. No my dear little man; always remember that the one true certain final and all important difference between you and an ape is that you have a hippopotamus major in your brain, and it has none; and that, therefore, to discover one in its brain will be a very wrong and dangerous thing, at which everyone will be

very much shocked. . . . Though really after all it don't much matter . . . if a hippopotamus was discovered in an ape's brain why it would not be one, you know, but something else.[84]

This passage not only lampooned the debate over the hippocampus, but also made light of the Wilberforce-Huxley clash, where Wilberforce supposedly asked Huxley whether he was related to an ape on his grandfather's or grandmother's side.

Both *A Sad Case* and *The Waterbabies* were insightful satires about the contemporary debates concerning evolutionary theory. Personality conflicts, fighting for power and prestige within the scientific community, and the clash with theology all played a role in the controversy over human origins. Huxley was at the center of these intrigues and certainly relished the chance to topple the "British Cuvier" from his pedestal. Yet in all of these accounts the technical aspects of the debate were portrayed as well. It is virtually impossible to disentangle the scientific disagreements from the larger social issues in which they are embedded. Still, following the chronology of Huxley's work between 1858 and 1863, it seems clear that the primary motive for his research on the relationship between apes and humans were the scientific issues involved. Huxley wanted the ape-human question to be regarded as a scientific one and to remove it from the realm of philosophy and theology. His involvement with this issue had begun before the publication of the *Origin*, but its publication further sparked Huxley's interest, resulting in a series of publications and public lectures. The investigation of these scientific questions brought him into conflict with Owen, not the other way around.

History as Told by the Winners

History has come down on the side of Huxley, regarding Owen as a nasty, pompous man whose science could not keep up with the times. Author Adrian Desmond has attempted to provide a more balanced account of the role Owen played in Victorian paleontology and his conflict with Huxley. He writes, "Treated sympathetically, Owen comes across as a psychologically complex character struggling with immensities—driven by his hatred of ape ancestry to search for a sensible non-transmutational answer to the biological challenge of the age."[85] Yet it seems that Owen succeeded in alienating just about everyone, not just Huxley. Even Desmond admits that while Huxley invariably brought out the worst in Owen,

> In the final analysis, Owen was notoriously insensitive, and occasionally callous. . . . Even continentals sympathetic to Owen's cause were put off by the man himself. The German zoologist Victor Carus confided to Huxley that for all his "admiration" of Owen, he "blushed" at his effrontery. Carus was keen to introduce Owen's osteological ideas into Germany. . . . [T]he fact that Carus ended up translating Darwin's *Origin of Species* and Huxley's *Man's Place in Nature* shows how Owen could alienate even potential allies.[86]

Rupke also thinks that Owen has been dealt with unfairly, claiming that Owen has been "systematically written out of Victorian history by Darwin and his followers."[87] He is quite correct that virtually all accounts of Owen's numerous disagreements with Huxley merely reflect Huxley's version of the story. But this imbalance is not due to some conspiracy on the part of the Darwinians.

Virtually all that is left of primary documents concerning the conflict between Huxley and Owen is what Huxley has written or correspondence to Huxley. The *Life and Letters of Richard Owen* were heavily edited by his grandson and contain only material that reflects favorably on Owen. A huge Owen correspondence resides in the British Museum of Natural History as well as the Royal College of Surgeons (RCS). Owen, like

Darwin, appeared to save and collect everything. However, significant gaps in his correspondence exist. Jacob Gruber, who has provided a detailed calendar of the correspondence, believes that significant editing of the letters occurred. It is obvious that letters have been pulled from the RCS collection because the numerical sequence of this collection has been interrupted.[88] Missing are three letters from 1853 and 1854 that may have been about Owen's difficulties at the RCS and that may imply he was thinking about emigrating to America. Three letters are missing from the British Association for the Advancement of Science (BAAS) meeting in Aberdeen in September 1859 when it was very likely that discussions occurred about the *Origin*, which was about to be published. Two letters from December 1860 and two from the period of the 1860 BAAS meetings at Oxford are missing. Most significantly, two letters are missing from the period of the Cambridge BAAS meetings when the real fight broke out between Owen and Huxley. Owen claimed that he would have nothing to do with Huxley, and Huxley was supposed to have called Owen a liar.[89] Gruber thinks this editing probably was done by Owen's grandson but does not rule out the possibility that Owen may have pulled letters himself.

Owen was a prolific correspondent, yet at the height of the Darwinian controversy between 1860 and 1863, few letters exist, and even fewer refer to the question of evolution. For example, the Duke of Argyll was also a prolific writer and was opposed to evolution. A continual exchange of letters occurs between the duke and Owen between 1854 and 1859, including a letter dated December 2, 1859, that said, "I have read Darwin with great interest." But the next letter from Argyll, which deprecates Darwin's theory, is dated February 27, 1863. It seems extremely unlikely that the two men would have stopped writing in this period, especially since they both opposed Darwin's theory. No letters from Wilberforce are included in the collection. However, Owen had provided the scientific arguments, corrected the page proofs, and added additional information for Wilberforce's bitter review of Darwin in the *Quarterly Review*. There is no correspondence that refers to Owen's own harsh review of the *Origin* in the *Edinburgh Review*. Letters from this time period written by Adam Sidgwick to Owen commenting on Owen's disagreements with Darwin are included in the *Life and Letters of Adam Sedgwick*, yet there is nothing in the Owen correspondence.[90] Because there is evidence that Owen later thought of himself as an evolutionist, "if not an originator of that notion,"[91] it is not out of the question to think that Owen removed the more serious evidence of his opposition to the theory in the years immediately following publication of the *Origin*. If true, it is an ironic twist of fate, because by removing all documents concerning the controversial aspects of his life, Owen virtually guaranteed that history was written from the vantage point of Huxley and other members of the Darwinian camp.

Man's Place in Nature *in the Context of the General Theory of Evolution*

I have argued that Huxley wrote *Man's Place in Nature* specifically to support Darwin's theory. It is true that it also served as a popular vehicle to attack Owen's faulty anatomy and the conclusions he drew from that anatomy. However, the grounds for that dispute were laid much earlier, and with Flower's public dissection of ape brains the issue had been resolved by 1861. Thus, we need to draw a distinction between Huxley's reasons for believing that humans had a pithecoid ancestry and how he chose to convince others that this was true. Huxley was a highly skilled polemicist, and he realized that just the facts of anatomy would not be enough to convince men of their ape ancestory. Nevertheless, the reason Huxley believed that humans were not an exception to Darwin's theory of the origin of species can be found in the facts of primate anatomy and physiology. Certainly

Huxley was naive in his belief that a clear distinction could be made between "facts" and the interpretation of them. The structures of the brain could not be as easily identified as either Huxley or Owen would have us believe. Even Huxley's good friend Charles Kingsley ridiculed Huxley's extreme empiricism in his portrayal of Professor Pttmilnsprts in *The Waterbabies*. In addition, Huxley controlled the boundaries of the discourse by framing the issues in particular ways, thus admitting some "facts" and not allowing others. However, this does not mean we have to look to social, economic, or political factors to explain why Huxley held the views he did.

While we do not need to turn to a social constructionist account to understand Huxley's view of "man's place in nature," to appreciate the significance of this episode we do have to place Huxley's argument in the context of the times. Huxley had a much more formidable agenda in writing *Man's Place* than just convincing "men they were monkeys." The hippocampus debate is an excellent example of how scientific discourse often operates at several different levels. *A Sad Case* is among other things a commentary on the political struggle for power occurring within the scientific community. Moreover, the ape-human controversy provided a wonderful opportunity for Huxley to show the dangers of mixing theology with science. This served his larger political agenda of making science rather than the Church the source of moral authority and power in society. Huxley's rhetorical strategy drew on his own anatomical work to advance a disciplinary argument, which at the same time resulted in personal notoriety and promoted his institutional standing. But this should not overshadow Huxley's appreciation of Darwin's theory as a scientific achievement. He did not think transmutation had been proven, but as he wrote Lyell, "I view it as a powerful instrument of research. Follow it out, and it will lead us somewhere; while the other notion is like all the modifications of 'final causation,' a barren virgin."[92]

Huxley was absolutely convinced that humans shared a

common ancestory with the apes and that there was nothing degrading in such a claim. In doing so he made a powerful argument for evolution, asserting that

> Mr. Darwin's hypothesis is not, so far . . . inconsistent with any known biological fact; on the contrary, if admitted, the facts of Development, of Comparative Anatomy, of Geographical Distribution, and of Paleontology, become connected together, and exhibit a meaning such as they never possessed before. . . . [Darwin's] hypothesis is as near an approximation to the truth as for example the Copernican hypothesis was to the true theory of planetary motions.[93]

Huxley was optimistic about the insights that evolution could provide to furthering human understanding. In the 1860s, he believed that the key to successfully playing the game of life was learning the rules of the game. For Huxley, the rules were the laws of nature. The game of life was infinitely more difficult and complicated than chess, and the other player was hidden from us, though his play was always "fair, just, and patient." Education consisted of learning the rules, and the teacher could only be Nature herself. If people directed their affections and wills "into an earnest and loving desire to move in harmony with [Nature's] laws," this would lead to a just and fair society.[94] Huxley believed that those who took "honors in Nature's University" were the really great and successful men in the world.

By the time of the Romanes lecture, however, Huxley's views had changed considerably. Herbert Spencer, who had coined the phrase "survival of the fittest," which Darwin later adopted to describe the ongoing struggle for existence resulting in natural selection, had articulated the advantages of applying evolutionary theory to social behavior, espousing an ethic that became known as "Social Darwinism." Spencer and his followers argued that one's moral obligations should be to promote this struggle for existence in the social realm. Thus, he was against any sort of safety net such as the poor laws, for they only contributed to the survival of the least fit.[95] Huxley could not abide such an ethic that was counter to all common decency, that claimed the state had no obligation to the less fortunate members of society. The Romanes lecture was written specifically in response to the extreme individualism and the harsh social policies Spencer was advocating in the name of evolution. In it Huxley claimed that

> laws and moral precepts are directed to the end of curbing the cosmic process and reminding the individual of his duty to the community. . . . Let us understand, once and for all that the ethical progress of society depends, not on imitating the cosmic process, still less in running away from it, but in combating it.[96]

Huxley, like many later critics such as G. E. Moore, attacked evolutionary ethics on the grounds of committing the naturalistic fallacy. Just because nature is a certain way does not mean nature *ought to be* that way. However, Huxley's critique actually goes far deeper than this.

Implicit in the various versions of evolutionary ethics was the idea that nature was progressive. Huxley denied this. For Huxley, one of the strengths of Darwin's theory was that in addition to explaining how organisms change and progress, it also explained how many organisms do not progress, and some even become simpler. Thus, why should we assume that applying the principles of evolution to the social realm would result in the progress and improvement of society? Huxley realized that "fittest" had a connotation of "best," but as he correctly pointed out, if the environment suddenly became much cooler, the survival of the fittest would most likely bring about in the plant world a population of more and more stunted and humbler organisms. In such an environment, the lichen and diatoms might be the most fit.[97] Furthermore, the strict definition of

Darwinian fitness is reproductive success. However, surely no one would label a mad rapist who successfully impregnates hundreds of women the "best" or "most fit" member of society.

Huxley may have been critical of evolutionary ethics, but Darwin's theory still had tremendous appeal because it was a purely naturalistic explanation of the history of life, free from any argument of design or supernatural causation. But in the best tradition of scientific inquiry, Huxley claimed he was not an advocate "if by an advocate is meant one whose business it is to smooth over real difficulties and to persuade when he cannot convince."[98] Thus, Huxley pointed out what he still saw as a significant problem of the theory: Artificial selection by plant and animal breeders had still not produced species which were incapable of interbreeding and thus threw into question the power of natural selection to create new species. Huxley's doubts about natural selection is the subject of the following chapter.

Notes

1. Thomas Huxley, March 22, 1861, *Life and Letters of T. H. Huxley* (*LLTHH*), 2 vols., Leonard Huxley, ed. (New York: D. Appleton & Co., 1900), 1: 205.

2. Charles Darwin, 1859, *The Origin of Species* (1859; New York: Avenel, 1976), p. 458.

3. Thomas Huxley, *Collected Essays of Thomas Huxley*, vol. 7, *Man's Place in Nature* (1863; New York: D. Appleton & Co., 1898), pp. 14–15.

4. See John Greene, *Death of Adam* (Ames: Iowa State University Press, 1959), pp. 178–79.

5. Huxley, *Man's Place*, p. 18.

6. Quoted in Greene, *Death of Adam*, p. 182.

7. Ibid., p. 184.

8. Quoted in E. L. Cloyd, *James Burnett, Lord Monboddo* (Oxford: Clarendon Press, 1972), p. 44.

9. Mario di Gregorio, *T. H. Huxley's Place in Natural Science* (New Haven, Conn.: Yale University Press, 1984), p. 152, but see entire discussion, pp. 129–53.

10. *LLTHH* 1: 60–61.

11. Ibid., p. 65.

12. Thomas Huxley, November 21, 1850, *LLTHH* 1: 68.

13. Thomas Huxley, November 9, 1851, *LLTHH* 1: 101–102.

14. Thomas Huxley to Nettie Heathorn, November 7, 1851, *LLTHH* 1: 75.

15. Thomas Huxley to Nettie Heathorn, April 14, 1851, *LLTHH* 1: 73.

16. Thomas Huxley quoting Richard Owen to Nettie Heathorn, May 4, 1851, *LLTHH* 1: 74.

17. Ibid.

18. Thomas Huxley to his sister Lizzie, May 20, 1851, *LLTHH* 1: 103.

19. Thomas Huxley, November 9, 1851, *LLTHH* 1: 102.

20. Thomas Huxley to Nettie Heathorn, May 4, 1851, *LLTHH* 1: 74.

21. Thomas Huxley to Lizzie, March 5, 1852, *LLTHH* 1: 105.

22. Thomas Huxley, November 9, 1851, *LLTHH* 1: 101.

23. Ibid.

24. Thomas Huxley to Edward Forbes, November 27, 1852, Huxley Manuscripts (HM) 16:172.

25. Thomas Huxley to Nettie, March 5, 1852, *LLTHH* 1: 106.

26. Ibid.

27. Thomas Huxley to John Churchill, July 22, 1857, HM 12: 194.

28. John Churchill, July 24, 1857, HM 12: 196.

29. Leonard Huxley, 1900, *LLTHH* 1: 153.

30. See chapter 2 for further discussion on this topic.

31. Thomas Huxley, "On the Theory of the Vertebrate Skull," *Proceedings of the Royal Society,* vol. 9 (1857–59), pp. 381–457, *Scientific Memoirs of Thomas Henry Huxley* (*SMTHH*), 4 vols, Michael Foster and E. Ray Lancaster, eds. (London: Macmillan & Co., 1898–1902), 1: 585.

32. Sir Michael Foster, 1895, Royal Society Obituary Notice of Huxley, quoted in *LLTHH* 1: 153.

33. Preface to 1894 edition of *Collected Essays of Thomas Huxley,* vol. 7, *Man's Place in Nature* (New York: D. Appleton & Co., 1984), p. viii.

34. Ibid.

35. Michael Bartholomew, "Huxley's Defense of Darwin," *Annals of Science* 32 (1975): 525–35.

36. Huxley, *Man's Place in Nature,* p. viii.

37. Quoted by Rev. W. H. Freemantle, "Account of the Oxford Meeting," 1892, *LLTHH* 1: 200.

38. Thomas Huxley, 1860, quoted by John Green in a letter to Boyd

Dawkins, *LLTHH* 1: 199. Oddly enough, no totally reliable version of what transpired at Oxford exists. Huxley, in a letter to Francis Darwin, claimed that Freemantle's account of the whole day was essentially correct, while Green had the substance of his speech most accurately, although he claimed that he definitely did not use the word "equivocal." While the accounts might differ in detail, in substance they are in agreement. Leonard Huxley has reprinted most of the different accounts of what happened, including Thomas Huxley's own comments on the various versions. See *LLTHH* 1: 193–204.

39. Leonard Huxley, 1900, *LLTHH* 1: 204.

40. Thomas Huxley, "On the Zoological Relations of Man with the Lower Animals," *Natural History Review* (1861): 67–84, *SMTHH* 2: 471–92.

41. Ibid., p. 474.

42. Ibid.

43. Ibid., pp. 471–72.

44. Ibid., p. 475.

45. Ibid., p. 472.

46. Ibid.

47. Thomas Huxley, 1893, "Evolution and Ethics," in *T. H. Huxley's "Evolution and Ethics": With New Essays on Its Victorian and Sociobiological Context*, James Paradis and George C. Williams, eds. (Princeton, N.J.: Princeton University Press, 1989), pp. 79–80.

48. Huxley, "On the Zoological Relations," *SMTHH* 2: 473.

49. Ibid. Huxley is quoting from Owen's "On Characters . . . &c., of the Class Mammalia," *Journal of the Proceedings of Linnaean Society of London* 2 (1857).

50. Ibid., p. 474, Thomas Huxley quoting Richard Owen, "On Characters," p. 33.

51. Ibid., p. 476, Huxley quoting Owen, p. 20.

52. Ibid., p. 476.

53. Ibid., Huxley quoting Allen Thomson's letter of May 24, 1860, pp. 481, 485.

54. Ibid.

55. Thomas Huxley to Joseph Hooker, April 18 and 27, 1861, *LLTHH* 1: 206.

56. Nicolaas Rupke, *Richard Owen, Victorian Naturalist* (New Haven, Conn.: Yale University Press, 1994), chapter 6.

57. Thomas Huxley, March 22, 1861, *LLTHH* 1: 205.

58. Leonard Huxley, 1900, *LLTHH* 1: 207.

59. Thomas Huxley, January 13, 1862, *LLTHH* 1: 209.

60. Thomas Huxley, January 16, 1862, *LLTHH* 1: 210.

61. Ibid.

62. Thomas Huxley, *Man's Place*, p. 81.

63. Ibid., p. 153.

64. Ibid., pp. 152–54.

65. Ibid., p. 154.

66. Ibid., pp. 155–56.

67. Ibid., p. 208.

68. Ibid.

69. Di Gregorio, *T. H. Huxley's Place*, p. 154.

70. Ibid.

71. Leonard Huxley, *LLTHH* 1: 217.

72. Charles Darwin to Charles Lyell, March 12, 1863, *Life and Letters of Charles Darwin* (*LLCD*), 2 vols., Francis Darwin, ed. (London: John Murray, 1881), 2: 99.

73. Charles Lyell, March 15, 1863, *Life and Letters of Charles Lyell*, 2 vols., Katherine Lyell, ed. (London: John Murray, 1881), 2: 366.

74. "Monkeyana," *Punch*, May 18, 1861, p. 206.

75. Anonymous, April 23, 1863, "A Sad Case, Mansion House," Owen Manuscripts, British Museum, Natural History Collection, p. 3.

76. Ibid., p. 6.

77. Ibid., p. 5.

78. Ibid., p. 6.

79. Ibid., p. 7.

80. Ibid.

81. Ibid., p. 8.

82. Charles Kingsley, *The Waterbabies* (1872; London: J. M. Dent & Sons, 1985), p. 50.

83. Ibid., pp. 107–109.

84. Ibid., pp. 109–10.

85. Adrian Desmond, *Archetypes and Ancestors* (Chicago: University of Chicago Press, 1984), p. 36.

86. Ibid., pp. 38–39.

87. Rupke, *Richard Owen*, p. 3.

88. Jacob Gruber, March 23, 1988, personal communication.

89. Ibid.

90. Jacob Gruber, "An Introductory Essay," Owen Collection (O.C.) 87, p. 18, British Museum, Natural History Collection.

91. Ibid.

92. Thomas Huxley, *LLTHH* 1: 187

93. Huxley, *Man's Place,* p. 149.

94. Thomas Huxley, "A Liberal Education and Where to Find It," *Collected Essays,* vol. 3, *Science and Education* (1868; New York: D. Appleton & Co., 1898), pp. 82–83.

95. See a series of twelve letters: "The Proper Sphere of Government," published in the *Nonconformist,* June–November 1842.

96. Thomas Huxley, *Collected Essays of Thomas Huxley,* vol. 9, *Evolution and Ethics* (London: Macmillan, 1894), pp. 82–83.

97. Ibid., pp. 80–81.

98. Huxley, *Man's Place,* p. 149.

Chapter Seven
Huxley and Natural Selection

Huxley's views about evolution "evolved" over time. Initially a saltationalist, he eventually adopted Darwin's gradualist position. In time, he also accepted the doctrine of progressive development after arguing against it for many years. However, he remained skeptical his entire life of the power of natural selection to create new species. What were his reasons, and can he still be called a Darwinian in spite of his reservations concerning the most basic tenet of Darwin's theory?[1]

In Huxley's famous *Times* review of the *Origin*, he stated that while Darwin's theory explained a great deal about the natural world, he personally preferred to adopt Goethe's aphorism "*Thatige Skepsis*," or active doubt.[2] He did not deny that natural selection existed in Nature, but could it account for all the effects that Darwin ascribed to it? Huxley, following Darwin, compared natural selection to artificial selection. He acknowledged that there appeared to be no limit to the amount of divergence that could be produced by artificial selection and that the races formed had a strong tendency to reproduce themselves. He admitted that "if certain breeds of dogs, or of pigeons, or of horses, were known only in a fossil state, no naturalist would hesitate in regarding them as a distinct species."[3] However, all these cases had been produced by human inter-

231

ference. "Without the breeder there would be no selection and without selection no race."[4] Darwin's "ingenious hypothesis" claimed that Nature took the place of the breeder. In the struggle for existence those individuals with variations that allowed them to compete most successfully survived in greater numbers, and through successive generations they eventually acquired the characters of a new species. Huxley believed that the Darwinian hypothesis provided an explanation for many apparent anomalies in the distribution of living beings in time and space. "That it is not contradicted by the main phenomena of life and organization appear to us to be unquestionable," he wrote.[5] But had Darwin been "led to over estimate the value of the principle of natural selection as greatly as Lamarck over estimated his *vera causa* of modification by exercise?"[6]

Huxley amplified his qualms about natural selection in his 1860 essay titled "The Origin of Species." He again posed the question to his readers: "Is it satisfactorily proved, in fact, that species may be originated by natural selection? that there is such a thing as natural selection?"[7] Until these questions could be answered in the affirmative, Darwin's proposal would remain a hypothesis and not a theory for Huxley, because "it is not absolutely proven that a group of animals having all the characters exhibited by species in Nature has ever been originated by selection, whether artificial or natural." Since species are reproductively isolated, then for natural selection to explain the origin of species, it has to be able to bring about that isolation. It wasn't clear to Huxley that natural selection actually caused reproductive isolation and therefore, it was perhaps an incomplete explanation for the origination of species. However, it was a hypothesis that was highly probable, "indeed the only extant hypothesis which is worth anything in a scientific point of view; but still a hypothesis, and not yet the theory of species."[8] Huxley's terminology is somewhat confusing here. To clarify, a hypothesis that had overwhelming evidence in the form of facts would, in his mind, move to the status of theory. Huxley also was overly confident as to what constituted "facts of nature." His problem with natural selection was twofold. First, he had different criteria than Darwin for what characteristics were crucial in defining a species, and second, he had a different view of what constituted proof of a hypothesis.

The Problem of Hybrid Sterility

Huxley's doubts about natural selection revolved around the question of hybrid sterility—the inability of distinct species to interbreed and produce fertile offspring. He observed that

> Groups having the morphological character of species, distinct and permanent races . . . have been so produced over and over again, but there is no positive evidence at present that any group of animals has by variation and selective breeding, given rise to another group which was in the least degree infertile with the first.[9]

For Huxley, proof of natural selection would have to be experimental proof, either observed in the wild or by means of artificial breeding experiments. He drew a distinction between morphological and physiological species. He wrote Kingsley that although Darwin had demonstrated "that selective breeding is a *vera causa* for morphological species; he has not yet shown it is a *vera causa* for physiological species."[10] Morphological species were not true species at all, according to Huxley. Rather, they were varieties or races that only looked like distinct species. Breeders had been quite successful in creating such "species." Pigeon fanciers, by artificial selection, had made distinct races such as the pouter, carrier, fantail, and tumbler. Huxley admitted, "No one would hesitate to describe the pouter and the tumbler as distinct species if they were found fossil[ized] . . . and without doubt if considered alone, they are good and distinct morphological species."[11] But they failed his criterion for physiological species in that the offspring between the races were perfectly fertile and were "descended from a common stock, the rock pigeon."[12] On the other hand, a cross between a horse and an ass resulted in the sterile donkey. A horse and an ass were physiological species, but how had they arisen? For Huxley, good, true physiological species were by definition incapable of interbreeding or produced offspring that were sterile. Thus, he drew a clear distinction between varieties or races that could still interbreed and species that could not.

British biologist and psychologist George Romanes also had a theory of physiological selection or selection for sterility factors. He claimed that natural selection did not account for the origin of species, but rather explained cumulative development of adaptation. Although his views are somewhat similar to Huxley's in certain respects, they are also quite different. Like Huxley, Romanes regarded hybrid sterility as a crucial criterion for defining new species. But Romanes also maintained that natural selection *could not* account for the origin of species. Huxley never claimed that. In fact, he believed that eventually proof would be obtained for natural selection; the proof just did not yet exist. Furthermore, there is no evidence that Romanes and Huxley discussed this matter, and Romanes developed his ideas much later.[13]

Darwin had a very different view of species, claiming that "there is no essential distinction between species and varieties."[14] Sterility was "not a specially acquired or endowed quality, but is incidental on other acquired differences."[15] Nevertheless, Darwin took Huxley's objections quite seriously and between 1860 and 1864 tried to convince Huxley that the sterility problem was not insurmountable. He agreed with Huxley that "the difficulty is great," but claimed that "it seems to me quite hopeless to attempt to explain why varieties are not sterile, until we know the precise cause of sterility in species."[16] Darwin asked Huxley to

Reflect for a moment on how small and on what very peculiar causes the unequal reciprocity of fertility in the same two species must depend. Reflect on the curious case of species more fertile with foreign pollen than their own. Reflect on many cases which could be given . . . of very slight changes of conditions causing one species to be quite sterile and not affecting a closely allied species. How profoundly ignorant we are on the intimate relation between conditions of life and impaired fertility in pure species! . . . The whole case seems to me far too mysterious to rest a valid attack on the theory of modification of species.[17]

Darwin continued to investigate this problem and for a while held a different view on the origin of sterility than what he had put forth in the *Origin*. He wrote Huxley that he was beginning to think that as a result of his research on *Primula* (primroses) that sterility was "an acquired or *selected* character—a view which I wish I had facts to maintain in the 'Origin.' "[18] It had been well known by breeders that cross-fertilization increased the vigor and fertility of offspring, while self-fertilization diminished them. Thus, Darwin reasoned that it would be an advantage for plants to develop a means of preventing self-fertilization. His research on *Primula* indicated that this was exactly what had happened. The primrose and cowslip each existed in two distinct forms. In one form the style was long and the stamens were short, while in other form the style was short and the stamens long. The two forms were distinct with no gradation between them and were much too regular and constant to be due to mere variability. Darwin did crosses of all possible combinations and concluded that heterostyly (differences in the style) appeared to be an adaptation to prevent self-fertilization. Having made the suggestion that self-sterility had a selective origin, Darwin tentatively asked whether cross-sterility might also have originated by selection. Numerous examples existed of partial sterility resulting from the crossing of varieties. If Darwin could demonstrate that sterility between varieties was being selected to preserve the integrity of incipient species, this would answer Huxley's objections. But Darwin admitted that "many great difficulties would remain, even if this view could be maintained."[19] The argument could go the other way as well. The fact that offspring from crosses between different varieties were often partially sterile while crosses between individuals that botanists claimed were distinct species sometimes resulted in fertile offspring was evidence that hybrid sterility was *not* a good criterion for defining species. If this was the case, there was no reason that sterility factors should be selected for.

Huxley tried to reassure Darwin that his concerns over the question of hybrid sterility did not detract from his belief in natural selection. In 1862 he wrote to Darwin,

> I have told my students that I entertain no doubt that twenty years' experiments on pigeons conducted by a skilled physiologist, instead of by a mere breeder, would give us physiological species sterile *inter se*, from a common stock (and in this, if I mistake not, I go further than you do yourself), and I have told them that when these experiments have been performed I shall consider your views to have a complete physical basis, and to stand on as firm ground as any physiological theory whatever. . . . I am constitutionally slow of adopting any theory that I must needs stick by when I have once gone in for it; but for these two years I have been gravitating towards your doctrines, and since the publication of your *primula* paper with accelerated velocity. By about this time next year I expect to have shot past you and to find you pitching into me for being more Darwinian than yourself.[20]

Darwin continued his research on hybrids, enlisting the aid of W. B. Tegetmeier, a poultry breeder and bee master, and John Scott, a young horticulturist, to do experiments to try and meet Huxley's demand for cross-sterile varieties that were produced by selection. Scott continued the work on *Primula* and reported finding "absolute zero fertility, apparently attained between undoubted varieties of a species!" He concluded, "In view of such evidence, I think I am fully justified in adding that this . . . form is, in fact judged by the physiological test so much insisted on by Professor Huxley, *a new and distinct species.*"[21] Now Darwin hopefully had evidence to meet Huxley's objections head on.

In spite of this new research, Huxley continued to bring up "his old line" about the hybrid sterility difficulty. In the fifth of a series of six public lectures to working men that were published in book form in 1863, he asked, "Can we find any approximation to this [hybrid sterility] in the different races known to be produced by selective breeding from a common stock? Up to the present time the answer to that question is absolutely a negative one. As far as we know at present there is nothing approximating to this check."[22] He again repeated the examples of the different races of pigeons that freely interbred together. Darwin felt that Huxley put far too much weight on animal results and was ignoring the plant breeding experiments. Exasperated, he wrote Huxley,

> You say the answer to varieties when crossed being at all sterile is "absolutely negative." Do you mean to say that Gärtner lied after experiments by the hundred . . . when he showed that this was the case with *Verbascum* and with maize . . . does Kölreuter lie when he speaks about the varieties of tobacco? My God, is not the case difficult enough, without its being, as I must think, falsely made more difficult?[23]

However, Huxley had not overlooked Darwin's arguments.

Even in his 1860 essay "The Origin of Species," Huxley acknowledged that there were varieties of plants when crossed whose offspring were almost sterile, while there were both animals and plants regarded by naturalists as distinct species that produced normal offspring when crossed. He cited Darwin's evidence that there were some plants that were more fertile when crossed with the pollen of another species than with their own. In addition, some plants, such as certain *Fuci*, gave the anomalous result that pollen of species A could successfully fertilize the ova of species B, but the reverse was not true—a cross between pollen from species B and ova from species A was ineffective. Huxley agreed with Darwin that "the sterility or fertility of crosses seems to bear no relation to the structural resemblances or differences of the members of any two groups."[24] He also realized there were many practical difficulties in applying his hybridization test. Many wild animals wouldn't breed in captivity, so the negative results of such crosses were meaningless. Since many plants were hermaphrodites, it was difficult to be positive that the plants when crossed had not been self-fertilized as well. In addition, experiments to determine the fertility of hybrids must be continued for a long time—particularly in the case of animals. He reiterated all of this evidence of partial sterility in the working man's lectures. Nevertheless, he claimed that "selective breeding can produce structural divergence as great as those of species, but we cannot produce equal physiological divergences."[25] In his final lecture he maintained,

> Mr. Darwin, in order to place his views beyond the reach of all possible assault, ought to be able to demonstrate the possibility of developing from a particular stock by selective breeding two forms, which should either be unable to cross one with another, or whose cross-bred offspring should be infertile with one another.[26]

Until the above criterion had been met, Huxley claimed that natural selection had not produced "all the phenomena which you have in nature. . . . I do not know that there is a single fact which would justify any one in saying that any degree of sterility has been observed between breeds absolutely known to have been produced by selective breeding from a common stock."[27]

Huxley did not deny that partial sterility existed between varieties, only that it had not been demonstrated to be the result of selection—either natural or artificial. He qualified his statement by saying that no facts existed that justified anyone asserting that such sterility could not be produced by proper experimentation and believed that eventually such sterility would be produced. But Darwin was not assuaged. Totally frustrated, he wrote Huxley,

> We differ so much that it is no use arguing. To get the degree of sterility you expect in recently formed varieties seems to me simply hopeless. It seems to me almost like those naturalists who declare they will never believe that one species turns into another till they see every state in the process.[28]

Darwin was right about Huxley. Until physiological speciation was demonstrated, Huxley would remain unconvinced of the efficacy of natural selection. Darwin knew that the type of evidence Huxley wanted was going to be difficult to find. In fact, some of his research indicated that speciation would *not* occur through the agency of artificial selection. He noted that in several cases two or three species had blended together and were now fertile. He wrote Huxley,

> Hence I conclude that there must be something in domestication—perhaps the less stable conditions, the very cause which induces so much variability which eliminates the natural sterility of species when crossed. If so we can see how unlikely that sterility should arise between domestic races.[29]

However, this was a problematical statement. Darwin's theory was based on the premise that a great amount of variation existed in nature. Natural selection acted on this variation, causing increasing divergence, which eventually led to speciation. On the other hand, he was claiming that there was increased variation with domestication, and whatever was causing it also somehow contributed to the lack of sterility between varieties, in spite of artificial selection. Huxley was quite correct that this was a weak point in Darwin's theory.

The dispute between Huxley and Darwin was further complicated because Darwin went back to his original position on the origin of hybrid sterility. In the above exchanges between the two men Darwin thought that Huxley was not giving enough weight to the evidence that did exist in favor of sterility between varieties and that he was being unrealistic to expect total infertility between varieties to be achieved within a few generations. But he agreed with Huxley that hybrid sterility was an important criterion for speciation. He was hopeful that he could demonstrate that sterility was being selected for. But further research on trimorphism in *Lythrum* convinced Darwin that his original view on the origin of sterility *was* correct in spite of his findings in the primrose.[30] In the sixth edition of the *Origin* he wrote that sterility "is an incidental result of differences in the reproductive systems of the parent-species."[31] Thus, Huxley and Darwin were as far apart in their views as when they began their dispute.

Huxley never moved from his original position. In the preface to *Darwiniana*, published two years before his death in 1893, Huxley wrote,

> Those who take the trouble to read the first two essays, published in 1859 and 1860, will, I think, do me the justice to admit that my zeal to secure fair play for Mr. Darwin, did not drive me into the position of a mere advocate; and that, while doing justice to the greatness of the argument I did not fail to indicate its weak points. I have never seen any reason for departing from the position which I took up in these two essays. . . . I remain of the opinion expressed in the second, that until selective breeding is definitely proved to give rise to varieties infertile with one another, the logical foundation of the theory of natural selection is incomplete. We still remain

very much in the dark about the causes of variation; the apparent inheritance of acquired characters in some cases; and the struggle for existence within the organism which probably lies at the bottom of both of these phenomena.[32]

Darwin did not believe that sterility was being selected for, but at the same time he claimed this did not undermine the power of natural selection to create new species. However, Huxley claimed that since true physiological species could not interbreed successfully, until such species had been produced either artificially or in the wild, the power of natural selection remained questionable. This was a more fundamental disagreement with Darwin than over the interpretation of experimental results. It was not that Huxley disagreed with Darwin's analysis of hybrid sterility, but rather that the two men had different expectations for what constituted proof of a hypothesis.

What Constitutes Proof of a Hypothesis?

Michael Ruse has pointed out that in the nineteenth century two different usages of the term *vera causa* existed: the empiricist and the rationalist *vera causa*. Darwin had been influenced by William Whewell, who based his *vera causa* on consilience. This was the rationalist *vera causa*.[33] Darwin argued not that there was direct experimental evidence for natural selection, but that it had wide explanatory power: "I must freely confess, the difficulties and objections are terrific, but I cannot believe that a false theory would explain as it seems to me it does explain so many classes of facts."[34] Huxley, however, wanted experimental proof. His *vera causa* was that of the empiricist. As Darwin wrote Hooker, Huxley "rates higher than I do the necessity of Natural Selection being shown to be a *vera causa* always in action."[35] Huxley acknowledged that fertility or infertility may be of little value as a test for speciation. Nevertheless, he maintained,

It must not be forgotten that the really important fact so far as the origin of species goes, is, that there are many things in Nature, as groups of animals and plants, the members of which are incapable of fertile union with those of other groups; that there are such things as hybrids, which are absolutely sterile when crossed, with other hybrids. . . . Such phenomena to which the name species is given, would have to be accounted for by any theory of the origin of species and every theory which couldn't account for it would be imperfect.[36]

For Huxley, the only way the theory of natural selection could definitely be proved was by empirical demonstration.

Darwin realized that Huxley was placing an essentially impossible demand on his theory. What was crucial for Darwin was not experimental demonstration of a hypothesis, but rather that the hypothesis could explain a great deal of observed phenomena. Darwin's *vera causa* was that of the rationalists. Thus, he wrote in *Animals and Plants under Domestication*:

The principle of natural selection may be looked at as a mere hypothesis, but rendered in some degree probable by what is positively known of the variability of organic beings in a state of nature—by what we positively know of the struggle for existence, and the consequent almost inevitable preservation of favourable variations—and from the analogical formation of domestic races. Now this hypothesis may be tested,—and this seems to me the only fair and legitimate manner of considering the whole question—by trying whether it explains several large and independent classes of facts; such as the geological succession of organic beings, their distribution in past and present times, and their mutual affinities and homologies. If the principle of natural selection does explain these and other large bodies of facts, it ought to be received.[37]

Darwin, of course, used empirical evidence for his theory, relying heavily on the analogy of artificial selection. But it was an analogy, and as it came under increasing attack, primarily from Huxley, Darwin fell back on a more general claim for his theory of natural selection: "I have always looked at the doctrine of natural selection as a hypothesis, which if it explained several large classes of facts, would deserve to be ranked as a theory deserving acceptance."[38]

Ruse has claimed that while Darwin moved freely back and forth between the empiricist and rationalist *vera causa*, relying on whichever one better suited his argument, Huxley always used the empiricist *vera causa*. Certainly, he prided himself on his empiricism. He believed the best way he could support Darwin's theory was to provide "facts" that were independent of theoretical considerations. However, he then argued that these facts made the most sense if interpreted within a Darwinian framework. It must be remembered that Huxley was attracted to Darwin's theory precisely because it did make sense of so many "facts of nature." He continually asserted that the facts of paleontology, of embryology, of taxonomy could best be explained by Darwin's theory. In Huxley's final lecture to working men he assessed the position of Darwin's work in relationship to a complete theory about the nature of the organic world. Huxley pointed out that while he had often quoted Darwin in previous lectures, it "has not been upon theoretical points, or for statements in any way connected with his particular speculations, but on matters of fact, brought forward by himself, or collected by himself, and which appear incidentally in his book."[39] Huxley recognized the great theoretical power of Darwin's theory, but it was precisely because it made sense of so many disparate facts of nature that he applauded it. Thus, he did make use of the rationalist *vera causa* in his defense of Darwin's general theory of evolution, but this is because the distinction between the two types of *vera causa* was not absolute. After all, consilience would have no force in the absence of evidence for the elements being brought together. Nevertheless, Huxley demanded that the empiricist *vera causa* be the criterion to determine whether natural selection be regarded as a theory or a hypothesis.

Why did Huxley have such a stringent requirement for natural selection when for the general theory of evolution he was willing to accept the idea of consilience? He often argued, just as Darwin did, that the theory explained large classes of facts in embryology, paleontology, and so forth. Perhaps the answer lies in his use of the term "physiological species."[40] Physiology was just becoming established as a true *experimental* science. Huxley had long advocated not only that natural sciences be taught in the schools, but also that the course should involve actual laboratory practice to train future scientists. In 1870, Huxley complained before the Royal Commission about the lack of facilities for practical teaching at the School of Mines, and finally after the passage of Elementary Education Bill of 1870, he was given an opportunity to organize such a course. It soon became quite famous, and the first courses in elementary biology given at Cambridge, Oxford, and Johns Hopkins universities were modeled on it.[41] In 1866 Trinity College was considering establishing a praelectorship in natural science, and Huxley urged that it be established in pure physiology, as distinguished from medicine.[42] He believed experimental physiology would be able to provide direct proof for natural selection by doing the type of experiments that would never be possible in paleontology. Again, it must be emphasized that he did not deny that natural selection could create new species, only that it had not been experimentally demonstrated or observed in the wild to be the case. But demanding such empirical evidence meant that for Huxley, natural selection would remain a hypothesis—a hypothesis that was highly probable, but nevertheless a hypothesis. Even today Huxley's criterion has not been met. How should we regard this disagreement between Huxley and Darwin in our assessment of Huxley's place in the history of evolutionary theory?

Huxley as an Advocate for Darwinism

As I noted in my introduction, the recent historiography on Huxley has become critical of Huxley's status as a Darwinian, primarily because of his lack of interest in natural selection and his belief in the type concept that was a reflection of his training as a developmental morphologist. I discussed in chapter 2 why developmental and evolutionary viewpoints are not antagonistic to one another. Huxley, in fact, represents the embodiment of both, and Darwin certainly borrowed heavily from the developmental tradition in building his theory. I also pointed out why no fundamental shift in thinking was necessary for taxonomy to incorporate Darwinian theory. Descent theory explained why organisms could be classified into related groups, but it did not provide a methodology for how to determine what those groups actually were. How does one actually ascertain who is descendent from whom? Huxley certainly had contradictory stances on the role of evolutionary theory in determining taxonomic relationships. He stated that classification should remain independent of phylogeny, and yet he became heavily involved in determining the phylogenies of a variety of groups and then used the phylogenies as a basis for classification. But regardless of his view on the importance of phylogeny for taxonomy, natural selection would not be relevant in determining either phylogenies or general classificatory schemes. Even today, natural selection plays no role in cladistic analysis. Thus, in Huxley's work on dinosaurs, on the relationship of birds to reptiles, and on horse phylogeny—research that was specifically concerned with bringing support for the theory of evolution—it should not be surprising that Huxley did not make use of natural selection. To assert that Huxley was a pre-Darwinian anatomist or a pseudo-Darwinian because he did not use natural selection in this type of work is simply false.[43]

Natural selection simply was not relevant to the types of questions that interested Huxley. As he candidly admitted of himself, "Not withstanding that natural science has been my proper business, I am afraid there is little of the genuine naturalist in me. I never collected anything and species work was always a burden to me. What I cared for was the architectural and engineering part of the business."[44] Should he have immediately started doing breeding experiments, looking at variation and effects of natural selection when he had never been particularly interested in those types of problems before? Precisely the sorts of interests Huxley denied having were the very ones that Alfred Wallace claimed were necessary to develop the theory of natural selection. Both Wallace and Darwin had the "mere passion of collecting" rather than the passion of

> studying the minutiae of structure either internal or external . . . an intense interest in the mere *variety* of things. . . . It is the superficial and almost child like interest in the outward forms of living things, which though often despised as unscientific happened to be *the only one* which would lead us towards a solution of the problems of species.[45]

For Wallace, it was the "constant search for and detection of these often unexpected differences between very similar creatures [that led] Darwin and myself . . . to thinking upon the 'why' and the 'how' of all this wonderful variety in nature."[46]

Although there were good reasons as to why Huxley would not use the principle of natural selection in his own research, this does not address the charge that Huxley cannot be considered a true Darwinian while harboring doubts over the power of natural selection. But as Huxley repeated many times in promulgating Darwin's theory, he was not an advocate, if an advocate meant ignoring what he perceived as difficulties in the theory. The question of hybrid sterility *was* a problem, and Darwin was the first to admit it. After all, no one had observed speciation in the wild. All the fantastic varieties that had been produced by artificial selection had still not resulted in the creation of actual new species. Rather than just pointing out that Huxley disagreed with Darwin over natural selection, a more sensitive assessment would emphasize the role Huxley played in helping Darwin to clarify his own views about the importance of sterility factors in speciation. As I will discuss in my final chapter, discussions over natural selection remain contentious among present-day evolutionary biologists and indeed demonstrate that the issues Huxley raised were not trivial.

As has been shown, Huxley's disagreement with Darwin over hybrid sterility was rooted more in philosophical differences over what constituted proof of a hypothesis than actual interpretation of experimental results. Darwin agreed with Huxley that if he could demonstrate that factors contributing to hybrid sterility were selected for, this would be powerful support for his theory. He conducted a series of experiments, enlisting the aid of other researchers to try and achieve this goal. While the research on *Primula* was extremely encouraging, the experiments with *Lythrum* resulted in Darwin returning to the initial view he presented in the *Origin*: sterility was not selected for but rather was the byproduct of other acquired traits. But Darwin's position was far stronger because of these experiments, and it is not at all clear that Darwin would have undertaken them if Huxley had not continually raised the issue. After Huxley's *Westminster* review of the *Origin* in 1860, Darwin met with Huxley and discussed the sterility problem at length. Darwin wrote Lyell, "I *think* I have convinced him [Huxley] that he has hardly allowed weight enough to the case of varieties of plants being in some degrees sterile."[47] However, Huxley was not convinced, much to Darwin's disappointment. He wrote Huxley,

> Some who went half an inch with me now go further, and some who were bitterly opposed are now less bitterly opposed. And this makes me feel a little disappointed that you are not inclined to think the general view in some slight

degree more probable than you did at first. This I consider rather ominous.[48]

Huxley's continued skepticism was probably key to Darwin's decision to continue experiments, enlisting the aid of Scott and Tegetmeier to try and demonstrate that the sterility difficulty was not overwhelming. Darwin ultimately referred to Huxley as the "Objector General" on the matter. Although these experiments failed to convince Huxley, did this seriously undermine his support for Darwin's theory?

In spite of his reservations about natural selection, Huxley was optimistic that evidence would eventually be brought to bear on the matter. He wrote Kingsley, "I entertain little doubt that a carefully devised system of experimentation would produce physiological speciation by selection—only the feat has not been performed yet."[49] In this respect Huxley was different from people such as the Duke of Argyll, Georges Mivart, George Romanes, and even Wallace, who argued not only that natural selection *did not* account for certain traits, but that it *could not* account for these traits. Huxley's doubts about natural selection did not prevent him from championing the cause of evolution. If one has to be in total agreement with every aspect of Darwin's theory, who would qualify as a Darwinian?[50]

Huxley recognized that the theory of evolution could be applied to a variety of different subjects, from embryology to paleontology. It provided a research program for a variety of different disciplines, ones that did not directly have much to do with natural selection. This continues to be true today. Molecular biology is a logical outgrowth of evolutionary theory. As Francis Crick wrote, the mechanism of natural selection is what

> makes biology different from all other sciences. . . . Natural selection almost always builds on what went before, so that a basically simple process becomes encumbered with many subsidiary gadgets. . . . It is the resulting complexity that makes

biological organisms so hard to unscramble. . . . Biological replication [is] central to the process of natural selection.[51]

Nevertheless, most molecular biologists are not concerned with natural selection. Huxley was a different sort of scientist than Darwin, interested in different types of problems. But this should not be interpreted to mean that he was not influenced by Darwin or that "Huxley became an evolutionist, but not of the Darwinian kind."[52] Most important for this discussion is that Huxley's criticism of natural selection did not prevent him from recognizing evolution as a theory with enormous explanatory power. An empiricist at heart, he was attracted to Darwin's theory because it explained a great deal of observed phenomena. He chose to defend Darwin by bringing the "facts" of evolution to the public, but this does not mean that he did not appreciate or understand the theoretical aspects of Darwinism. As he wrote Darwin after delivering his speech "On the Coming of the Age of the *Origin of Species*" at the Royal Institution April 9, 1880, "I hope you do not imagine because I had nothing to say about "natural selection" that I am at all weak of faith on that article. . . . But the first thing seems to me to be to drive the fact of evolution into people's heads; when that is once safe the rest will come easy."[53] As was mentioned in chapter 3, unlike Lyell, Huxley immediately saw the difference between Darwin's theory and Lamarck's, and that difference specifically revolved around natural selection. He wrote Lyell that Lamarck had no notion of modification by variation.

> If Darwin is right—about natural selection—the discovery of this *vera causa* sets him to my mind in a different region altogether from all his predecessors—and I should no more call his doctrine a modification of Lamarck's than I should call the Newtonian theory of the celestial motions a modification of the Ptolemaic system.[54]

A more powerful comparison would be hard to find, demonstrating that Huxley fully understood the ramifications of natural selection.

Convincing Victorians of their ape ancestry was a truly monumental task. For the general public as well as for most scientists, it was the most controversial and, therefore, fundamental aspect of Darwin's theory, more important than natural selection, gradualism or any of the many technical arguments put forth in the *Origin*. Huxley recognized this immediately. Where was God in the *Origin*, the public asked? Huxley devoted much of his public defenses of Darwinism to showing that this was not an appropriate question to ask of a scientific theory. At the risk of being accused of doing Whig history, I think the fact that creationism threatens the teaching of evolution in the United States, that many Americans do not believe in evolution, indicates that Huxley was correct in which aspects of Darwinism he chose to emphasize. However, this emphasis also reflected Huxley's deeply felt need for a totally naturalistic explanation of earth history. The next chapter examines Huxley's philosophical beliefs and how they influenced his defense of evolution.

Notes

1. The term "Darwinian" is problematic, and what it means has been actively debated among historians and philosophers of biology. However, the current challenge to Huxley's status as a Darwinian revolves primarily around his views about natural selection. See part 4, "Perspectives on Darwin and Darwinism," and especially David Hull's essay "Darwinism as a Historical Entity: A Historiographic Proposal," in *The Darwinian Heritage*, David Kohn, ed. (Princeton, N.J.: Princeton University Press, 1985), for a further discussion of these issues.

2. Thomas Huxley, 1859, "The Darwinian Hypothesis," in *Collected Essays of Thomas Huxley*, vol. 2, *Darwiniana* (London: Macmillan & Co., 1893), p. 20.

3. Ibid., p. 17.

4. Ibid.

5. Ibid., p. 20.

6. Ibid.

7. Thomas Huxley, 1860, "The Origin of Species," in *Darwiniana*, p. 74.

8. Ibid.

9. Ibid., pp. 74–75.

10. Thomas Huxley to Charles Kingsley, April 30, 1863, *Life and Letters of T. H. Huxley* (*LLTHH*), 2 vols., Leonard Huxley, ed. (New York: D. Appleton & Co., 1900), 1: 257.

11. Huxley, "Origin of Species," p. 44.

12. Ibid.

13. George Romanes, "Physiological Selection; an Additional Suggestion on the Origin of Species," *Journal of the Linnean Society of London* (Zoo.) 19 (1886): 337–411.

14. Charles Darwin, *The Origin of Species* (1859; New York: Avenel, 1976), p. 288.

15. Ibid., p. 264.

16. Charles Darwin to Thomas Huxley, January 11 [1860?], *More Letters of Charles Darwin* (*MLCD*), 2 vols., Francis Darwin and A. C. Seward, eds. (London: John Murray, 1903), 1: 137.

17. Ibid., pp. 137–38.

18. Charles Darwin to Thomas Huxley, January 14, 1862, *Life and Letters of Charles Darwin* (*LLCD*), 2 vols., Francis Darwin, ed. (London: John Murray, 1887), 2: 177.

19. Ibid.

20. Thomas Huxley to Charles Darwin, January 20, 1862, *LLTHH* 1: 211.

21. John Scott, "Observation on the Functions and Structure of the Reproductive Organs in the *Primulaceae*," *Journal of the Proceedings of the Linnean Society of London* (Bot.) 8 (1864): 97–108. Quoted in Malcolm Kottler, "Charles Darwin and Alfred Russel Wallace: Two Decades of Debate Over Natural Selection," in Kohn, *The Darwinian Heritage*, pp. 365–432.

22. Thomas Huxley, *On Our Knowledge of the Causes of the Phenomena of Organic Nature* (London: Robert Harwick, 1863), p. 112.

23. Charles Darwin, December 18, 1862, *MLCD* 1: 230.

24. Huxley, "Origin of Species," p. 47.

25. Huxley, *On Our Knowledge*, p. 116.

26. Ibid., p. 146.

27. Ibid., pp. 146–47.

28. Charles Darwin to Thomas Huxley, December 28, 1862, *MLCD* 1: 225.

29. Charles Darwin to Thomas Huxley, January 10, 1863, *MLCD* 1: 232.

30. For a more complete discussion on Darwin's views on hybrid sterility see Kottler, "Charles Darwin and Alfred Russel Wallace." This essay also points out that Wallace, like Huxley, also found the problem of hybrid sterility a serious one.

31. Charles Darwin, *The Origin of Species*, 6th ed. (London: John Murray, 1872; New York: Collier Books, 1962), p. 275.

32. Thomas Huxley, 1893, preface to *Darwiniana*, pp. v–vi.

33. Michael Ruse, *The Darwinian Revolution* (Chicago: University of Chicago Press, 1979, 1981), pp. 235–36.

34. Charles Darwin to Hugh Falconer, December 17, 1859, *MLCD* 1: 455.

35. Charles Darwin, February 14, 1860, *MLCD* 1: 140.

36. Huxley, "Origin of Species," p. 49.

37. Charles Darwin, *Animals and Plants under Domestication*, 2d ed., 2 vols. (New York: D. Appleton & Co., 1872, 1892), 1: 9.

38. Charles Darwin to Joseph D. Hooker, February 14, 1860, *MLCD* 1: 140.

39. Huxley, *On Our Knowledge*, p. 132.

40. I thank Marc Swetlitz for suggesting this idea to me.

41. For a good discussion of Huxley's role in the rise of physiology see Gerald Geison, *Michael Foster and the Cambridge School of Physiology* (Princeton, N.J.: Princeton University Press, 1978).

42. Ibid., p. 76.

43. See Michael Bartholomew, "Huxley's Defense of Darwin," *Annals of Science* 32 (1975): 525–35, Michael Ghiselin, "The Individual in the Darwinian Revolution," *New Literary History* 3 (1971): 113–34, and Peter Bowler, *The Non-Darwinian Revolution* (Baltimore: Johns Hopkins University Press, 1988), who take such a position.

44. Thomas Huxley, "Autobiography," in *Collected Essays of Thomas Huxley*, vol. 1, *Methods and Results* (1889; New York: D. Appleton & Co., 1893), p. 7.

45. Alfred Wallace, *Darwin–Wallace Celebration Held on Thursday, 1st July, 1908 by the Linnean Society of London* (London: The Linnean Society, 1908), p. 8.

46. Ibid., p. 9.

47. Charles Darwin, April 10, 1860, *LLCD* 2: 94.

48. Charles Darwin, December 2, 1860, *LLCD* 2: 147.

49. Thomas Huxley to Charles Kingsley, April 30, 1863, *LLTHH* 1: 257.

50. As David Hull points out, one needs to make a "distinction between the Darwinians as a social group and Darwinism as a conceptual system. A scientist can be a Darwinian without accepting all or even a large proportion of the elements of Darwinism," in "Darwinism as a Historical Entity: A Historiographic Proposal," in Kohn, *The Darwinian Heritage*, p. 809.

51. Francis Crick, *What Mad Pursuit* (New York: Basic Books, 1988), p. 5.

52. Mario di Gregorio, "The Dinosaur Connection: A Reinterpretation of T. H. Huxley's Evolutionary View," *Journal of the History of Biology* 15 (1982): 417.

53. Thomas Huxley, May 10, 1880, *LLTHH* 2: 13.

54. Thomas Huxley, August 17, 1862, Huxley Manuscripts (HM) 30: 41.

Chapter Eight

Evolution and Huxley's Worldview

Two Paths to Knowledge?

Huxley, in his later years, wrote that the goal for schoolmen as well as for ourselves should be to settle "the question how far the universe is the manifestation of a rational order; in other words, how far logical deduction from indisputable premises will account for that which has happened and does happen."[1] He claimed that medieval scholasticism shared this goal with modern science: "Modern science takes into account all the phenomena of the universe which are brought to our knowledge by observation and experiment."[2] Huxley acknowledged that two worlds existed—the physical and the psychical. Although the two were intimately related and interconnected, "the bridge from one to the other has yet to be found."[3] For Huxley, science and philosophy both contributed to our knowledge of the world, but they represented distinct approaches toward that knowledge. He articulated the distinction between these two approaches in his essay on Descartes's *Discourse on Method.*[4] For Huxley, Descartes stood at the branch point between philosophy and science of the modern world. One branch led by way of Berkeley and Hume to Kant and idealism, while the other, by way of Julien De La Mettrie and Joseph

Priestley, led to modern physiology and materialism. Both pathways shared the stipulation to "give unqualified assent to no propositions, but those the truth of which is so clear and distinct that they cannot be doubted."[5] Descartes's "consecrated doubt" provided Huxley with a guiding principle for his own work and underlay his agnosticism. Like Goethe's *Thatige Skeptsis*, or active skepticism, this skepticism was not the type "born of flippancy and ignorance, and whose aim is only to perpetuate itself as an excuse for idleness and indifference."[6] Huxley's doubt, like Descartes's, differed from the skeptics, "who doubt only for doubting's sake, and pretend to be always undecided; on the contrary, my whole intention was to arrive at a certainty, and to dig away the drift and the sand until I reached the rock or the clay beneath."[7]

According to Huxley, "Each of the two branches [physics and metaphysics] were sound and healthy and has as much life and vigor as the other." He asserted that metaphysics and physics were complementary, not antagonistic, and claimed that "thought will never be completely fruitful until the one unites with the other. . . . Descartes's two paths meet at the summit of the mountain, though they set out on opposite sides of it."[8] But physics and metaphysics could be reconciled only if each side acknowledged its own faults. The materialists must acknowledge that all our knowledge of nature was known to us only by the facts of consciousness, and the metaphysician had to admit that progress in the understanding of consciousness could be made only by the methods and formulae of science.

Huxley may have claimed that both branches were necessary and each were "sound and healthy," but in actuality physics made up the bulk of his tree of knowledge. At best, metaphysics was only a minor twig, and in many of his writings the twig was diseased. Better to cut the twig off because it endangered the growth of the whole tree. Huxley's idiosyncratic definition of materialism split it between both branches. He placed materialistic terminology on the branch of physics, while materialist philosophy went on the twig of metaphysics.

Huxley was often called a materialist, which he firmly denied. However, it is easy to understand why he was labeled one: Although he made a distinction between materialist terminology and materialist philosophy, he advocated using materialistic terminology to describe the world. While criticizing the philosophical materialists, he agreed with them that

> the human body, like all living bodies, is a machine, all the operations of which will, sooner or later, be explained on physical principles. I believe that we shall, sooner or later, arrive at a mechanical equivalent of consciousness, just as we have arrived at a mechanical equivalent of heat.[9]

As an example, he suggested that

> If a pound weight falling through a distance of a foot gives rise to a definite amount of heat, which may properly be said to be its equivalent; the same pound weight falling through a foot on a man's hand gives rise to a definite amount of feeling, which might with equal propriety, be said to be its equivalent.[10]

Huxley commented again and again that materialist methodology had led to tremendous advances in physiology and psychology, but that materialist philosophy involved "grave philosophical error."[11] He had no use for the philosophical materialists who claimed that there was nothing in the universe but matter, force, and necessary laws because he asserted that

> all our knowledge is a knowledge of states of consciousness. "Matter" and "Force" are, as far as we can know, mere names for certain forms of consciousness. "Necessary" means that of which we cannot conceive the contrary. "Law" means a rule which we have always found to hold good, and which we expect always will hold good.[12]

Thus, the material world was known to us only by the forms of the ideal world. In making such a claim, Huxley followed Hume rather than Descartes. For Descartes matter was substance that had extension but did not think, while spirit was substance that thinks but had no extension. Huxley, however, found Descartes's phraseology confusing, especially since Descartes located the soul in the pineal gland. Huxley interpreted this to mean that the soul was a mathematical point, having place (the pineal gland) but not extension. It also exerted force since it had free will to alter the course of the animal spirits, which were nothing more than matter in motion. Therefore, the distinction between matter and spirit disappeared, since matter in this analysis might be nothing more than centers of force.[13] Scientists no longer believed that the soul was lodged in the pineal gland, asserting that the seat of consciousness was located in the gray matter of the cerebrum. But as Huxley correctly pointed out, this only made matters worse in trying to understand the nature of consciousness using Descartes's spirit/matter distinction. The brain, being material, had extension, and if the soul was lodged in it, it must also have extension. Once again spirit was lost in matter. Still, Huxley did not deny the existence of the spiritual world, but like Hume, he maintained that at the present time we had no way of having any knowledge of it.

Huxley was drawing a subtle distinction between materialist terminology and materialist philosophy. And although he wrote extensively on the distinction, on the limitations of philosophical materialism, nevertheless Huxley followed the materialist path in his own personal quest for knowledge. His 1868 lecture "On the Physical Basis of Life" epitomized a materialistic worldview and threw the religious community into an uproar.

Appearing before a large Edinburgh audience with a bottle of smelling salts, water, and various other common substances, Huxley declared he had all the basic ingredients of protoplasm, or what he translated as the "physical basis of life." Matter and life were inseparably connected: "There is some one kind of matter which is common to all living beings."[14] The lichen, the pine, the fig, a Finner whale, the flower in a girl's hair, and the girl all exhibited a threefold unity: "a unity of power, a unity of form, and a unity of substantial composition."[15] By unity of power, Huxley meant that all the activities of living organisms, from amoebas to humans, fell into three basic categories. They were involved in maintenance and development, or they effected changes in the relative position of parts of the body, or they were involved in continuance of the species.

> Even those manifestations of intellect, of feeling, and of will, which we rightly name the higher faculties, are not excluded from this classification. . . . Speech, gesture, and every other form of human action are, in the long run resolvable into muscular contraction, and muscular contraction is but a transitory change in the relative positions of the parts of a muscle.[16]

All living organisms shared a common unity of form—the cell—and all cells shared a similar chemical composition. Plants and animals appeared to be very different, yet no sharp dividing line existed between the simplest of these organisms. Even the distinction between living and nonliving matter lay in the arrangement of molecules.

> All vital action may . . . be said to be the result of the molecular forces of the protoplasm which displays it. And if so, it must be true, in the same sense and to the same extent, that the thoughts to which I am now giving utterance, and your thoughts regarding them, are the expression of molecular changes in the matter of life which is the source of our other vital phenomena.[17]

After delivering such a discourse, is it any wonder that people called Huxley a materialist?

To be fair, "The Physical Basis of Life" was not just an expli-
cation of materialism. Anticipating that he would be accused of
"gross and brutal materialism," Huxley attempted to show why
materialism involved "grave philosophical error." Following
Hume and relying heavily on Spencer, Huxley asked,

> What do we know of this terrible "matter," except as a name
> for the unknown and hypothetical cause of states of our own
> consciousness? And what do we know of that "spirit" . . .
> except that it is also a name for an unknown and hypothetical
> cause, or condition, of states of consciousness? In other
> words, matter and spirit are but names for the imaginary sub-
> strata of groups of natural phenomena.[18]

Huxley touched briefly on the problem of necessity. One
could claim that it is a law of nature that an unsupported stone
will fall to the ground, it was quite another to say it *must* fall to
the ground. Empirical observation could not unequivocally
prove such a claim. Huxley believed that the fundamental doctrines of philosophical materialism, like those of spiritualism, lay
outside the "limits of philosophical inquiry," and Hume had provided a clear demonstration of what those limits were. Huxley's
discussion of Hume expressed his own agnosticism in a nutshell.
   Although Hume called himself a skeptic, Huxley believed
this did not do justice to the power and subtlety of Hume's
intellect. Many questions were not worth being skeptical about.
If someone asked Huxley what the politics of the inhabitants of
the moon were, he was not being skeptical when he said he
didn't know. No one could possibly know the answer to such a
question. Hume pointed out many interesting problems that
pique our curiosity, but he had demonstrated that they were
essentially questions of lunar politics. Huxley asked, "Why
trouble ourselves about matters of which however important
they may be, we do know nothing, and can know nothing?"[19]
However, he did not dismiss such inquiries just because they

were a waste of time. For Huxley the investigation of nature served a higher purpose than merely satisfying our curiosity. In a world full of misery and ignorance, he believed every person had a duty to try to leave it "somewhat less miserable and somewhat less ignorant than it was before he entered it."[20]

Huxley claimed that it was of little moment whether matter was regarded as a form of thought or thought was regarded as a form of matter. Each statement had a certain relative truth. However, he preferred the terminology of materialism because it provided a method of inquiry into the phenomena of nature, including the nature of thought, by studying physical conditions that were accessible to us. He believed that the "spiritualist terminology is utterly barren and leads to nothing, but obscurity and confusion of ideas."[21] Exactly what did Huxley mean when he used the word "terminology"? He was referring not just to language, but to a method of inquiry. Many would claim that a method of inquiry is, in fact, a philosophy. Furthermore, by describing spiritualist terminology as "barren" and leading to a "confusion of ideas," he certainly implied that spiritualistic metaphysics itself was barren. Materialism might not answer everything, but for Huxley, spiritualistic metaphysics answered nothing, even if it posed interesting questions. Speculative philosophy could only point out the limits of knowledge. Huxley truly may have been an agnostic on many important questions. However, if he believed that the scientific method provided the only useful way of increasing knowledge, he should not have been surprised that he was accused of being a materialist.

Huxley attacked Auguste Comte in another attempt to distance himself from the materialists. He thought that Comte knew nothing about physical science, claiming that Comte's "classification of Sciences is bosh."[22] Huxley had no use for someone who regarded phrenology as a great science and psychology a chimera, and who described Cuvier as "brilliant, but superficial."[23] Anything that was positive in the positivism of

Comte could be found in the earlier writings of philosopher David Hume.[24] More important, Huxley believed that Comtism was in spirit antiscientific. He claimed that Comte's philosophy was "antagonistic to the very essence of science as anything in ultramontane Catholicism" and described Comte's philosophy as "Catholicism *minus* Christianity."[25] In two sentences Huxley had succeeded in offending Comtians and Catholics alike. But Huxley had essentially condensed and paraphrased what Comte had written in the fifth volume of the "Philosophie Positive": "Comte's ideal as stated by himself is Catholic organization without Catholic doctrine or in other words, Catholicism *minus* Christianity."[26]

In his 1826 essay "Considérations sur le Pouvior spirituel," Comte advocated the establishment of a "modern spiritual Power," which he hoped might have even greater influence over temporal affairs than the Catholic clergy at the height of their power. This spiritual power would have control over public opinion and education. Furthermore, Comte saw no point in doubting scientific principles that had been established by competent persons. For Huxley, "nothing in ultramontane Catholicism" could be more antiscientific than what Comte had proposed in this essay and, therefore, nothing more offensive. Progress had been made only by just those people who had not been hesitant to doubt established beliefs. "The great teaching of science—the great use of it as an instrument of mental discipline—is its constant inculcation of the maxim, that the sole ground on which any statement has a right to be believed is the impossibility of refuting it."[27] Thus, the discussion had come full circle, returning to Descartes's guiding principle of doubt.

Huxley wrote in despair to Darwin that people continually distorted his views on materialism and agnosticism. "I begin to understand your sufferings over the *Origin*. A good book is comparable to a piece of meat, and fools are as flies who swarm to it, each for the purpose of depositing and hatching his own particular maggot of an idea."[28] "Materialist," "positivist," and even "atheist" were labels often applied to him, but "agnostic" is the best epithet for him. "My fundamental axiom of speculative philosophy is that materialism and spiritualism are opposite poles of the same absurdity—the absurdity of imagining that we know anything about either spirit or matter."[29] As the author of his obituary notice in the *Times* wrote, "Nothing could be more unjust to a man of so absolutely skeptical a mind as Huxley than to charge him with anything so rashly positive as Atheism."[30]

Huxley's Antitheology and His Defense of Evolution

One may quibble whether Huxley was a materialist or not, but his anticlericalism cannot be doubted. Comte's philosophy was permeated by the "papal spirit," thus ensuring a caustic attack by Huxley. Rather than questioning authority, Catholic doctrine demanded acceptance of ideas based solely on authority. What could be more fundamentally opposed to the spirit of free scientific inquiry? For Huxley there could be "neither peace nor truce" between agnosticism and clericalism.

> The Cleric asserts that it is morally wrong not to believe certain propositions, whatever the results of a strict scientific investigation of the evidence of these propositions. He tells us "that religious error is, in itself, of an immoral nature." He declares that he has prejudged certain conclusions, and looks upon those who show cause for arrest of judgment as emissaries of Satan. It necessarily follows that, for him, the attainment of faith, not the ascertainment of truth, is the highest aim of mental life. And, on careful analysis of the nature of this faith, it will too often be found to be, not the mystic process of unity with the Divine, understood, by the religious

enthusiast, but that which the candid simplicity of a Sunday scholar once defined it to be.[31]

Blind acceptance of authority was never acceptable to Huxley.

Huxley did not rule out the possibility of genuine religious experience. However, he objected to Catholicism and most organized religions because of the authority structure they relied on to foster belief. Cardinal Newman had defined faith as "the power of saying you believe things which are incredible." For Huxley, faith described in this way was an "abomination." Huxley's war was with theology rather than with religion. Theology asked people to believe statements about the nature of the world without any evidence or, worse yet, when the evidence clearly contradicted church teachings. Contributing a chapter entitled "On the Reception of the *Origin of Species*" for *The Life and Letters of Charles Darwin*, he wrote:

> I had not then, and I have not now, the smallest *a priori* objections to raise to the account of the creation of animals and plants given in "Paradise Lost," in which Milton so vividly embodies the natural sense of Genesis. *Far be it from me to say that it is untrue because it is impossible* [emphasis added]. I confine myself to what must be regarded as a modest and reasonable request for some particle of evidence that the existing species of animals and plants did originate in that way, as a condition of my belief in a statement which appears to me to be highly improbable.[32]

Sarcasm dripping from his pen, he asked, "Where was the evidence for the story of Genesis?" Huxley wanted to understand the natural world. Theology obscured that understanding, while Darwinism went a long way in elucidating it.

For the most part, Huxley's anticlerical views and his desire to keep theology totally separate from science served him well in his defense of Darwinism. We have seen how he used it as a strategy to convince people of the validity of evolution in his

American lectures. Such a separation also played a crucial role in his attempt to convince "men that they were monkeys." Many have claimed that the elimination of a theologically grounded view of science is the true legacy of Darwinism. Huxley, I'm sure, would ascribe to such a view. John Greene has argued that a Darwinian worldview inevitably led toward agnosticism and mechanistic materialism.[33] Richard Lewontin, Steven Rose, and Leon Kamin agree, as does Ernst Mayr.[34] The most extreme proponent of this position is William Provine, who claims that the Darwinian revolution will be complete when we all become atheists.[35] Taking exception to this view of history, Robert Richards acknowledges that "Huxley's materialistic epiphenomenalism" was one research program that logically followed from a Darwinian worldview, but asserts that the monism of Romanes, Morgan, James, and Baldwin traces its roots back to Darwinism as well.[36] Perhaps, but Huxley realized that theological objections were at the root of most criticisms of Darwinian theory, no matter how sophisticated the arguments appeared to be scientifically. Charles Lyell could never accept the evolution of man from a lower form, ultimately because of his religious views regarding the dignity of man. Richard Owen also regarded evolution as a threat to the special status of man. As a result, not only did he initially argue against Darwin's theory, but he made the fallacious claim that only humans possessed a hippocampus. By referring to natural selection as the law of higgledy-piggledy, Sir John Herschel implied that the complexity of living organisms could not have arisen by such a chancy mechanism. Even people who supported evolution often distorted Darwin's view in order to reconcile it with a belief that God still played a role in shaping the history of life on earth. People such as Asa Gray and the Duke of Argyll reformulated Darwin's mechanism of natural selection by arguing that the Creator somehow guided the type of variations produced for natural selection to act on or otherwise exerted some type of control over the evolutionary process. For instance, the Duke of Argyll claimed that the magnificent coloring of some birds could not have been the result of utilitarian adaptation. He dismissed the possibility that such plumage played a crucial role in attracting a mate. Rather, the Creator must have a sense of beauty which he expressed by guiding the evolutionary process along certain lines. In other words, both the Duke of Argyll and Asa Gray cast evolution within the old idealist version of design.[37]

In *Post Darwinian Controversies*, James Moore examined the struggles of Christians to come to terms with Darwin.[38] He discusses not only the Christian anti-Darwinians, the group most often subject to Huxley's caustic attacks, but also two groups of Christians who wanted to support evolution. According to Moore, Christian Darwinians "found room for Darwin's science in a fresh understanding of the historic doctrine of a triune God."[39] Another group, whom he called Christian Darwinists, reconciled Darwinism and Christian doctrine by making use of non-Darwinian evolutionary theories such as Lamarckian evolution.[40] In Moore's analysis, Huxley is severely criticized for lumping the variety of Christian responses to Darwinism into a single one that was misguided and rooted in ignorance. "In the polemical world of T. H. Huxley liberal reconcilers of Christianity and evolution could be nothing but an 'army' bent on blending scientific truth with theological error."[41] Certainly, Huxley oversimplified the kinds of discussions that were going on among the Christian Darwinists in his war against theology. William Gladstone, as a defender of orthodox Christianity, had claimed in an article in *Nineteenth Century* that geology had shown the account of creation in Genesis was scientifically correct. Huxley was infuriated that Gladstone presented himself as an authority on both science and religion and responded in the next issue by demonstrating that the geological and Genesis accounts of creation could not be reconciled. Furthermore, Judaism and Christianity would have to show evidence for their historical claims, which to Huxley's mind they had failed to do.[42] Nevertheless, Moore's analysis also demonstrates that the objections by both the anti-Darwinians and the Christian reconcilers of Darwinism were rooted in theology. As Moore wrote,

> Christian Darwinists therefore came into conflict with Darwinism because they believed that God's purposes are manifested in the world and that these purposes disclose God's omnipotence and beneficent character because, more precisely, they believed in a God whose purposes could not have been realized through evolution as Darwin conceived it. . . . They adopted theories of evolution which, by altering and adulterating Darwinism, were congenial to the purposes and character of God.[43]

An outstanding example of one who tried to reconcile Darwin's theories with Christianity was St. George Mivart, who accepted evolution but believed that it was constrained along certain lines by the Creator. Furthermore, Mivart also claimed that evolution was compatible with the teachings of the Church, and this ensured that he would be subjected to Huxley's caustic pen.

Mivart had received much of his scientific training under Huxley and in most biological matters followed Huxley, for whom he had profound respect. Thus, Mivart's later views caused Huxley much pain. As Jacob Gruber wrote of Mivart,

> This was not a Wilberforce or a Gladstone or any of the petty snipers in the religious press whose theological rumblings barely concealed their scientific ignorance. This was a colleague whose competence was a product of his own tuition. This was a friend and associate who had been warmly embraced by the small company of Darwinians.[44]

Mivart's objections to Darwinism, Huxley's response, and how the exchange was regarded by other members of the Darwinian camp demonstrate that Huxley was correct: The battle for evolution had to be fought on theological grounds. Mivart raised a series of scientific objections as to why natural selection could not account for the origin of species. Yet Huxley, for the most part, chose to ignore Mivart's scientific objections to Darwin's theory. Instead, in his review of Mivart's *Genesis of Species*, he devoted his energies to attacking Mivart's claim that Darwinism was compatible with the teachings of the Catholic Church.[45] Huxley appeared far more interested in promulgating his anti-Catholicism than defending Darwinism. But why did Huxley's critique of Mivart take the form that it did? I think that his response to Mivart was exactly in the spirit of Mivart's attack on Darwin.

In "Mr. Darwin's Critics," Huxley assailed Mivart's *Genesis of Species*, an anonymous review of Darwin's *Descent of Man* in the *Quarterly Review*, which Huxley correctly recognized as being written by Mivart, and Wallace's "Contributions to the Theory of Natural Selection." By 1871 Huxley believed that the quality of the anti-Darwinian criticism had improved significantly. No longer could the attacks be characterized as a "mixture of ignorance and insolence."[46] In particular, the critiques by Wallace and Mivart demanded a response, not only because of the authors' scientific competence, but also "because they exhibit an attention to those philosophical questions which underlie all physical science, which is as rare as it is needful."[47] Thus, Huxley was suggesting that the philosophical issues, rather than the scientific objections the two men had raised, were what needed to be addressed. Both Mivart and Wallace were "as stout believers in evolution as Mr. Darwin himself," but both men denied that man could have evolved from lower forms by the process of natural selection. Huxley claimed that Mivart was less of a Darwinian than Wallace because he had less faith in the power of natural selection for even the evolution of lower forms. But Mivart was more of an evolutionist than Wallace because Wallace believed that some kind of intelligent agent was necessary, "a sort of supernatural Sir John Sebright*—to

---

*John Sebright was an expert pigeon breeder whose research Darwin had cited in the *Origin*.

even produce the animal frame of man; while Mr. Mivart requires no Divine assistance till he comes to man's soul."[48]

In *Genesis* Mivart attempted to reach a compromise between science and religion. He had two objectives, first "to show that natural selection is not *the* origin of species. This was and is my conviction purely as a man of science, and I maintain it upon scientific grounds only." The second objective was "to demonstrate that nothing even in Mr. Darwin's theory, as then put forth and *a fortieori* in evolution generally, was necessarily antagonistic to Christianity."[49] But rather than deal with the scientific objections that Mivart raised, Huxley chose to devote over half his review to demolishing Mivart's latter objective. And Huxley quite clearly stated why he did so: "In addition to the truth of the doctrine of evolution, indeed, one of its greatest merits in my eyes, is the fact that it occupies a position of complete and irreconcilable antagonism to that vigorous and consistent enemy of the highest intellectual, moral, and social life of mankind—the Catholic Church."

Mivart wanted to demonstrate the compatibility of Catholic theology with evolution. By citing various orthodox Catholic sources, Mivart attempted to show that nothing in Church writings prevented an acceptance of evolution. He even suggested that their writings left open the possibility that evolution offered a reasonable explanation for species variation. One such Church authority Mivart cited was the Jesuit Suarez. Huxley made the teachings of Father Suarez the cornerstone of his attack on Mivart. He wrote Hooker that as a result of reading Mivart's book,

> the devil has tempted me to follow up his very cocky capsuling of Catholic theology based as he says upon Father Suarez. . . . Master Mivart either gushes without reading or reads without understanding. Frater Suarez would have damned him 40 times over for holding the views he does. . . . [Mivart] allows himself to be insolent to Darwin and I mean

to pin him out. Only fancy my vindicating Catholic orthodoxy against the Papishes themselves.[50]

Whereas Huxley briefly mentioned the writings of Augustine
and Aquinas, he discussed Suarez at great length, quoting long
passages in Latin. Huxley concluded, "These passages leave no
doubt that this great doctor of the Catholic Church, of unchal-
lenged authority and unspotted orthodoxy . . . declares it to be
Catholic doctrine that the work of creation took place in the
space of six natural days"[51] But as Huxley pointed out, the belief
that the universe was created in six natural days was hopelessly
inconsistent with the doctrine of evolution. Huxley continued
his discussion, noting that according to Suarez God made a soul
by direct creation and that woman was made out of the rib of
man. Furthermore, Suarez did not hesitate to criticize Augustine or Aquinas when their writings seemed to go against this
strict Catholic orthodoxy. Huxley's review thus far was entirely
devoted to demonstrating that the biblical account of Genesis
was totally incompatible with the teachings of evolution.

Even if Huxley was correct about Suarez, and Mivart was
wrong, Suarez played a relatively minor role in Mivart's entire
argument. More important, after thirty pages Huxley still had
not addressed Mivart's scientific objections. Some of these
objections to Darwin's theory were ones that Huxley also had,
for example the importance of saltations. Huxley had never hes-
itated to state his criticisms of Darwin's theory. But he also
pointed out the many reasons why he still supported Darwinism. This two-pronged strategy had been quite powerful.
Why did he change his tactics in this later review? I think the
answer is twofold. First, Huxley wanted to argue that theology
had no place in the practice of science, and second, he thought
that most objections to Darwinism were based on theological
concerns. Mivart had been Darwin's most persistent critic.
Thus, if Huxley could show that even Mivart's objections were
ultimately rooted in his theological beliefs, this would provide

powerful support to Huxley's claim that most objections to Darwin's theory were not scientific, but rather theological.

As we have seen in many of his other defenses of Darwin, Huxley was arguing not just that the theory of evolution provided a better description of the organic world than the teachings of the Church, but rather that science and theology were fundamentally incompatible. After spending some twenty pages presenting the teachings of the Catholic Church, he concluded,

> Thus far the contradiction between Catholic verity and Scientific verity is complete and absolute, *quite independently of the truth or falsehood of the doctrine of evolution*. But for those who hold the doctrine of evolution, all the Catholic verities about the creation of living beings must be no less false. . . . If Suarez has rightly stated Catholic doctrine, then evolution is utter heresy (emphasis added).[52]

But evolution, according to Huxley, was not heresy. Rather, it provided the best, closest approximation to truth we had about the workings of the natural world. "For me the doctrine of evolution is no speculation, but a generalisation of certain facts, which may be observed by anyone who will take the necessary trouble."[53] Evolution squarely contradicted Church teachings. For Huxley, all other points paled in significance. Both Darwin and Huxley had dealt with scientific objections to the theory in a variety of places. Moreover, Mivart had misrepresented Darwin's arguments in order to make his own case for the argument from design. Other Darwinians shared Huxley's perception as to the fundamental objection Mivart had to Darwin's theory. Darwin, John Tyndall, and Joseph Hooker all applauded the emphasis of Huxley's attack.

Darwin felt that Mivart had severely distorted his arguments as well as presented him as "the most arrogant, odious beast that ever lived."[54] He couldn't understand why Mivart had such animosity toward him, but supposed "that accursed religious bigotry is at the root of it."[55] Darwin was thrilled that Huxley, once again, was willing to defend him and wrote Huxley, "What a wonderful man you are to grapple with those old metaphysico-divinity books."[56] Darwin recognized that because of the theological objections, acceptance of his theory would be a long time in coming. "It will be a long battle, after we are dead and gone, as we may infer from Malthus even yet not being understood."[57] After reading the proofs of Huxley's soon-to-be-published critique of Mivart, Darwin wrote Huxley,

> How you do smash Mivart's theology; it is almost equal to your article versus Comte,—that never can be transcended. ... Nothing will hurt him so much as this part of your review. But I have been preeminently glad to read your discussion on his metaphysics, especially about reason and his exposition of it. I felt sure he was wrong, but having only common observation and sense to trust to, I did not know what to say in my 2nd Edit. of Descent. Now a footnote and a reference to you will do the work.[58]

Echoing Darwin's praise, Tyndall wrote: "Mivart you have handled most admirably. The remarks about his 'Guidance' are excellent and will give satisfaction to all good men."[59] Hooker also strongly supported the emphasis in Huxley's review:

> I have just glanced again at Mivart's last chapter; it is curious for the illustrations it adduces pro and con his views, which seem to have been sought with zeal and produced without discretion. The pages on the attributes of an Almighty God are hopelessly vague and common place. And I never had much respect for the God who originates *derivatively.* His "God inscrutable" is no better or worse for me than Spencer's "God unknowable" who he won't have! . . . The whole scheme of "Derivative Creation" in its religious aspects always seemed to me a poor makeshift—a sweet to the physic of evolution; and I should indeed be astonished if the Jesuit

father's conceptions of creation squared with this. All they contended for I assume, was that God made beasts and birds out of solids and not out of vacuum.[60]

Huxley, as the most outspoken member of the Darwinian camp, led the attack on theology and clericalism. But the personal exchanges between these men clearly document that they shared Huxley's views on what should be the focus of his attack. While most of "Mr. Darwin's Critics" criticized Mivart's theology, Huxley did devote some time to other issues. He discussed both Mivart's and Wallace's objections to the evolution of the human intellect and showed that their reasoning was faulty. He also pointed out some actual errors in Mivart's representation of Darwin's views. Yet rather than applauding Huxley for pointing out Mivart's misrepresentation, Darwin, Hooker, and Tyndall all praised Huxley's attack on Mivart's theology.

Huxley was correct. Mivart's objections *were* ultimately rooted in his theological beliefs, a fact that becomes apparent at the end of his review of *Descent of Man*: "A great part of the work may be dismissed as beside the point—as a mere elaborate and profuse statement of the obvious fact, which no one denies, that man is an animal."[61] But Darwin had not demonstrated that the difference between man and brute was only one of degree rather than kind. According to Mivart, man was not merely an intelligent animal. He was a free moral agent. This distinction was

> so profound that none of those which separate other visible beings is comparable with it. The gulf which lies between his being as a whole, and that of the highest brute, marks off vastly more than a mere kingdom of material beings; and man, so considered, differs far more from an elephant or a gorilla than do these from the dust of the earth on which they tread. Thus . . . the author of "The Descent of Man" has utterly failed in the only part of his work which is really important. Mr. Darwin's errors are mainly due to a radically false metaphysical system.[62]

Mivart made Darwin sound like a fool, and the Darwinians were not going to take this lightly.[63] We cannot dismiss the personal aspect of this dispute. Proper respect had not been shown to Darwin. Snide and disrespectful in tone, Mivart accused Darwin of presenting evidence "with a dogmatism little worthy of a philosopher."[64] He was surprised that Darwin accepted a particular tale "without suspicion" and asked if "Mr. Darwin had ever tested this alleged fact."[65] And as a final insult, he claimed Darwin had not "maturely reflected" over the data he collected. Nevertheless, Mivart's objections, like Owen's and Lyell's, revolved around the brute nature of man. Man was not just an animal. He was in a separate category, not just because of his intelligence, but because of his other attributes. Mivart asked his readers, "Must we acknowledge that man with all his noble qualities, with sympathy which feels for the most debased, with benevolence which extends . . . to the humblest living creature, with his god-like intellect . . . with all these exalted powers is descended from an Ascidian?"[66] Mivart attributed a special status to man because of his religious indoctrination. Thus, the thrust of Huxley's attack was on theology, although he also addressed Mivart's attack on Darwin.

The conflicts surrounding Darwinism were complex and cannot be reduced to an analysis that just describes science at war with religion. Just as today, many people in Huxley's time found a way to reconcile evolution with their religious beliefs.[67] However, it was not those people Huxley was concerned with. Rather, he was concerned with the opponents of evolution, and he believed that the vast majority of them rejected Darwinism because of their religious beliefs, religion being broadly conceived to include more than just the concept of a Christian God. I do not deny that Huxley pitted evolution against theology to promote his larger agenda, which was to replace the power and moral authority of the Church with that of the church of science. However, Huxley correctly recognized that the battle for the acceptance of evolution had to be fought on theological grounds, not just in the public arena, but within the scientific community as well.

In my first chapter, I claimed that Huxley's philosophical views provided the framework for his scientific views. I think it could also be argued that his experience as a scientist provided the framework for his philosophy. Science was the path to knowledge, and as Huxley stated again and again, this was his goal in life—to better understand the natural world. It is important to stress that the mid–nineteenth century was a particularly exciting time for biology. The cell theory of Schleiden and Schwann, embryological research, improvements in microscopy, the truly dramatic finds in paleontology, and an enormous increase in data on the distribution of plants and animals all provided evidence for Darwin's theory that earlier theories about the history of life lacked. This does not lessen Darwin's accomplishment in any way, but it does speak to Huxley's claim that nature provided overwhelming evidence in support of Darwin's theory: "evolution is no longer a speculation, but a statement of historical fact."[68] Darwin's theory had tremendous appeal for Huxley because it was a purely naturalistic explanation of the history of life, free from any argument of design, or supernatural causation. While it is true that Huxley continually used Darwin's theory as a vehicle to attack theology, this should not overshadow Huxley's appreciation of the theory as a scientific theory—as a complete theory about the nature of the organic world. In spite of his emphasis on "the facts," Huxley also recognized the great theoretical power of the *Origin*, claiming that there was "no field of biological inquiry in which the influence of the 'Origin of Species' is not traceable" and that "as the embodiment of an hypothesis, it is destined to be the guide of biological and psychological speculation for the next three or four generations."[69] How prophetic were Huxley's words. My final chapter examines some of the recent developments in modern biology that resonate with Huxley's interest in developmental morphology as it relates to evolutionary theory.

Notes

1. Thomas Huxley, 1887, "Scientific and Pseudo-Scientific Realism," in *Collected Essays of Thomas Huxley*, vol. 5, *Science and the Christian Tradition* (New York: D. Appleton & Co., 1898), p. 62.

2. Ibid.

3. Ibid.

4. Thomas Huxley, "On Descartes' 'Discourse Touching the Method of Using One's Reason Rightly and of Seeking Scientific Truth,' " *Collected Essays of Thomas Huxley*, vol. 1, *Methods and Results* (London: Macmillan & Co., 1893), pp. 166–98.

5. Ibid., p. 169.

6. Ibid., p. 170.

7. Ibid.

8. Ibid., pp. 191, 194.

9. Ibid., p. 191.

10. Ibid. Huxley also referred his reader to a discussion of the nature of the relation between nerve-action and consciousness in Herbert Spencer's *Principles of Psychology*.

11. Thomas Huxley, 1868, "On the Physical Basis of Life," in *Methods and Results*, p. 155.

12. Ibid., p. 193.

13. Ibid., p. 189.

14. Ibid., p. 131.

15. Ibid., p. 133.

16. Ibid., pp. 133–34.

17. Ibid., p. 154.

18. Ibid., p. 160.

19. Ibid., p. 163.

20. Ibid.

21. Ibid., p. 164.

22. Thomas Huxley to Charles Kinglsey, April 12, 1869, *Life and Letters of T. H. Huxley* (*LLTHH*), 2 vols., Leonard Huxley, ed. (New York: D. Appleton & Co., 1900), 1: 323.

23. Thomas Huxley, "The Scientific Aspects of Positivism," in *Lay Sermons, Essays, and Reviews* (1869; London: Macmillan & Co., 1899), p. 135.

24. Huxley, "On the Physical Basis of Life," pp. 156–58.

25. Ibid., p. 156.

26. Huxley, "The Scientific Aspects of Positivism," p. 133.

27. Ibid., p. 149.

28. Thomas Huxley, March 17, 1869, *LLTHH* 1: 323.

29. Thomas Huxley to Charles Kingsley, May 22, 1863, *LLTHH* 1: 262.

30. Obituary notice, August 5, 1895, *The Times*, Huxley Manuscripts (HM) 81:1:3.

31. Huxley, "Agnosticism and Christianity," in *Science and the Christian Tradition*, p. 313.

32. Thomas Huxley, 1887, "On the Reception of *The Origin of Species*," *Life and Letters of Charles Darwin* (*LLCD*), 2 vols., Francis Darwin, ed. (London: John Murray, 1887), 1: 541.

33. John Greene, "Darwinism as World View," in *Science, Ideology, and World View* (Berkeley: University of California Press, 1981).

34. Richard Lewontin, Steven Rose, and Leon Kamin, *Not in Our Genes* (New York: Pantheon, 1984), p. 51; Ernst Mayr, "The Nature of the Darwinian Revolution," in *Evolution and the Diversity of Life* (Cambridge: Harvard University Press, 1976), pp. 277–96.

35. William Provine, "Progress in Evolution and the Meaning of Life," in *Evolutionary Progress*, Matthew Nitecki, ed. (Chicago: University of Chicago Press, 1988), pp. 49–76.

36. Robert Richards, *Darwin and the Emergence of Evolutionary Theories of Mind and Behavior* (Chicago: University of Chicago Press, 1987), p. 407.

37. For a more thorough discussion of this point see Peter Bowler, "Darwin and the Argument from Design," *Journal of the History of Biology* 10 (1977): 29–43.

38. James R. Moore, *The Post-Darwinian Controversies* (Cambridge: Cambridge University Press, 1979).

39. Ibid., p. 252.

40. Ibid., p. 218.

41. Ibid., p. 217.

42. See, for instance, Thomas Huxley, 1887, "Science and Pseudo-Science," in *Science and the Christian Tradition*, pp. 115–16.

43. Moore, *Post-Darwinian Controversies*, pp. 218–19.

44. Jacob Gruber, *A Conscience in Conflict* (New York: Columbia University Press, 1960), p. 37.

45. Huxley, 1871, "Mr. Darwin's Critics," in *Collected Essays of Thomas Huxley*, vol. 2, *Darwiniana* (London: Macmillan & Co., 1893), pp. 120–86.

46. Ibid., p. 121.

47. Ibid.

48. Ibid., p. 122.

49. St. George Mivart, "Evolution and Its Consequences: A Reply to Professor Huxley," *Contemporary Review* 19 (1872): 168–97. Quoted in Gruber, *A Conscience in Conflict*, p. 53.

50. Thomas Huxley to Joseph Hooker, September 11, 1871, HM 2: 181.

51. Huxley, "Mr. Darwin's Critics," p. 138.

52. Ibid., pp. 146–47.

53. Thomas Huxley, *Science and the Christian Tradition*, p. 42.

54. Charles Darwin to Joseph Hooker, Sept 16, 1871, *More Letters of Charles Darwin (MLCD)*, 2 vols., Francis Darwin and A. C. Seward, eds. (London: John Murray, 1903), 1: 333.

55. Ibid.

56. Charles Darwin to Thomas Huxley, September 21, 1871, *LLCD* 2: 327.

57. Ibid., p. 328.

58. Charles Darwin, September 30, 1871, *LLCD* 2: 328–29.

59. John Tyndall, April 2, 1873, HM 1: 83.

60. Joseph Hooker, September 17, 1871, HM 3: 150.

61. St. George J. Mivart, "Darwin's Descent of Man," *Quarterly Review* 131 (July 1871): 47–90. In *Darwin and His Critics*, David Hull, ed. (Chicago: University of Chicago Press, 1983), pp. 354–84.

62. Ibid., p. 383.

63. For a good discussion of the personal aspect of this conflict see Gruber, *A Conscience in Conflict*, chapters 5 and 6.

64. Mivart, "Darwin's Descent of Man," p. 360.

65. Ibid., p. 366.

66. Ibid., p. 360.

67. For a good treatment of the subject see D. Lindberg and R. Numbers, eds., *God and Nature* (Berkeley: University of California Press, 1986). Chapters 11–15 are particularly relevant.

68. Thomas Huxley, 1880, "The Coming of Age of *The Origin of Species*," in *Darwiniana*, p. 242.

69. Ibid., p. 228.

Chapter Nine

Huxley's Biology and Modern Evolutionary Theory

"The doctrine that every natural group is organized after a definite archetype is . . . a doctrine which seems to me as important for zoology as the theory of definite proportions for chemistry."

Thomas Huxley, 1853

"Modern Biologists . . . exhibit a positive dread of form."

Marjorie Grene, 1974[1]

"My objective . . . is to develop a theory of biological form. . . . [Morphogenetic fields] are domains of spatial order, defined by internal relationships, that change in time according to well-defined principles or rules."

Brian Goodwin, 1996[2]

Development and Evolution

Huxley appointed himself "Darwin's bulldog," and a splendid bulldog he was. Paradoxically, such a legacy has meant that in spite of all that has been written on Huxley, his own research has not received the attention it deserves. The recent round of scholarship correctly places Huxley in a morphological tradition but then pits this interest against natural selection. Michael Bartholomew typifies how many people have regarded Huxley's own biological work. He writes that Huxley's scientific work "underwent no radical change in 1859," noting that *Man's Place in Nature* made no contribution to natural selection theory and that Huxley's scientific papers showed no interest in problems of variation, selection, or inheritance. His failure in this respect "must be set against his monumental achievement in introducing and habituating Victorians to the idea that they were descended from apes." Bartholomew concludes that Huxley's defense of Darwin "was encumbered with material which he had carried with him from an earlier, and proleptically anti-Darwinian phase."[3] One might ask how someone can be anti-Darwinian before Darwin's ideas were known, but more importantly, it is precisely because Huxley's research was in the tradition of developmental morphology with a commitment to the type concept, while still accepting evolutionary theory, unlike most other morphologists, that his views are so interesting.[4] Today, many of the issues that Huxley raised are providing the basis for what promises to be a profound revision in modern evolutionary theory. Research in development and paleontology

> have once again brought to the force evidence of regularity or constraint in evolution and in development, so that the search for laws of form and morphological transformation at the level of the whole organism has become a legitimate and exciting research program. . . . The evolutionary process would then become intelligible in terms of laws intrinsic to developmental processes as well as contingencies arising from extrinsic or environmental influences.[5]

Certainly, Huxley knew nothing about and could not have anticipated the developments in genetics, thermodynamics, or the rise of the "science of complexity." Nevertheless, his desire to discover the laws of form, his early saltational views, and his doubts about the efficacy of natural selection resonate with the concerns of many of today's most provocative and innovative researchers in evolutionary biology.

Darwin's theory is one of the most successful theories to have emerged in science. However, as Brian Goodwin has pointed out, scientific theories are the result of choices and assumptions that are neither arbitrary nor inevitable. All theories carry a particular viewpoint and in doing so bring clarity to specific aspects of reality, delegating other aspects to the background.[6] Darwin, heavily influenced (at least initially) by natural theology, accepted that the most important phenomenon of life that needed to be explained was the adaptation of organisms to their environment. Natural selection acting on random hereditary variation over long periods of time provided a mechanism that explained not just how species change, but how organisms change *adaptively*. Explaining adaptation has been Darwin's legacy, dominating the research agenda of evolutionary theory. For many people, modern evolutionary theory has come to mean essentially what is explained by natural selection, and that in turn has become the basis for explaining virtually all aspects of life on earth. "Evolution" and "Darwinism" are often regarded as synonymous terms, where evolution is defined as a "change in gene frequencies." Ironically, in such a definition, organisms, which Darwin regarded as the primary example of living nature, "have faded away to the point where they no longer exist as fundamental and irreducible units of life. Organisms have been replaced by genes and their products as the basic elements of biological reality."[7] Not only is such a gene-centered definition clearly not applicable to the nineteenth century, but it severely limits the kinds of questions evolutionary theory can address in the twentieth and twenty-first centuries.

Huxley had a much broader conception of the word "evolution," which is most clearly seen in his 1878 *Encyclopedia Britannica* article "Evolution in Biology." He pointed out that

in the eighteenth century "evolution" was virtually synonymous with the word "development." It "is in fact at present
employed in biology as a general name for the history of the
steps by which any living being has acquired the morphological
and the physiological characters which distinguish it."[8] Drawing
on the ideas of Descartes and elaborated by Spencer, Huxley
gave a much more general account of what the word "evolution" meant. All objects in the physical universe, whether living
or not, "have originated by a process of evolution, due to the
continuous operation of purely physical causes, out of a relatively formless matter."[9] Wallace and Darwin in their theory of
evolution by natural selection have confined themselves to addressing the problem of biological "evolution"—the causes
which have brought the present condition of living matter—
"assuming such matter to have come into existence." Huxley
certainly did not underestimate the ideas of Darwin and Wallace, asserting that "*The Origin of Species* was responsible for the
doctrine of evolution assuming a position of importance which
it never before possessed."[10] However, he was drawn to
Spencer's more encompassing definition of evolution: "a transformation of the homogeneous into the heterogeneous, the
indefinite into the definite, or the transformation of the inco-
herent into the coherent." Such a definition is very close to von
Baer's law of embryonic development, and thus we can see why
Huxley would find it attractive. For Spencer the "law of evolution" applied to inorganic processes, whether it be the formation of the solar system or the shaping of the earth. It applied
to the development of life on earth as well as the development
of society. Thus, biological evolution was just one aspect of this
more general law of evolution, which referred to a universal
process of spontaneous ordering, or self-organization. Spencer
never was able to demonstrate the physical basis for this general
law. However, modern research in thermodynamics and the
origin of life along with the rise of complexity theory have more
in common with Spencer's ideas than with the theory of natural

selection.[11] Stuart Kauffman has shown that complex systems of many kinds exhibit high spontaneous order and argues that this order may "enable, guide and *limit* selection. . . . We must invent a new theory of evolution which encompasses the marriage of selection and self-organization."[12] It is ironic that in Huxley's unrelenting proselytizing of Darwin's theory he helped contribute to evolution becoming reduced in meaning to that which can be explained by natural selection, a view that he did not hold himself. Such a narrowing meant that the question of form that so interested him became relegated to the background.

Although Huxley agreed with Spencer that biological evolution was just a special case of a more general ordering process, he still championed Darwin's theory because it had great explanatory power. He maintained that it had achieved its present high status because the accumulating facts in a variety of disciplines from embryology to geology to paleontology made a great deal of sense with the assumption of evolution and were virtually unintelligible without out it. As he pointed out in the *Britannica* article, the great *echelle des être* of Bonnet simply did not stand up in light of this new information. A much more logical classification grouped organisms into distinct types, and the resemblance and differences within the groups made sense if one interpreted them as branches that sprang from a common hypothetical ancestor. Furthermore, that hypothetical common ancestor often was assuming a tangible reality with new fossils being discovered almost daily. Many researchers were attempting to build phylogenetic trees based on homologues of one group with adult structures of another. Haeckel with his biogenetic law tried to show how evolution could occur by changes in embryonic development. Thus, the evidence for descent with modification was overwhelming, and that aspect of evolution was quite rapidly accepted. The same could not be said for natural selection.

"How far 'natural selection' suffices for the production of

species remains to be seen." Nevertheless, Huxley admitted that if it is not the "whole cause it must play a great part in sorting out of varieties into those which are transitory and those which are permanent." As Huxley pointed out and Darwin freely admitted, no well-grounded theory of heredity existed. Yet, inherited variation was the raw material for Darwin's theory. In the Austrian Alps, a monk by the name of Gregor Mendel was analyzing his crosses of pea plants in relative obscurity, and a science of genetics would have to wait another forty years. Since "the causes and conditions of variation have yet to be thoroughly explored," Huxley suggested that variation might be restricted along definite directions. It certainly was conceivable "that every species tends to produce varieties of a limited number and kind and that the effect of natural selection is to favour the development of some of these, while it opposed those of others along their predetermined lines of modification."[13] Both the grouping of present-day organisms and the gaps in the fossil record suggested that variation was not continuous. If we take these findings along with the work of von Baer, Cuvier, and Huxley, it was quite reasonable for Huxley to initially advocate a saltationalist view of evolutionary change. Although evidence from both paleontology and development resulted in Huxley eventually abandoning his saltational views, the questions that Huxley raised concerning both saltation and the power of natural selection have not been entirely resolved. Indeed, they are getting a rigorous rehearing.

Microevolution, Macroevolution, and the Fossil Record

Saltation in various guises has resurfaced periodically throughout the twentieth century and reflects deeper issues that evolutionary theory has failed to adequately explain. The rediscovery and independent verification of Mendel's work eventually vindicated Darwin's theory of natural selection. Ironically, Mendelian genetics was initially used to argue against natural selection. Hugo de Vries claimed that species-level change could occur only by natural selection acting on large mutations. De Vries's theory gave primacy of place to the mutation itself rather than natural selection in the actual cause of speciation. However, the rise of population genetics in the 1920s and 1930s demonstrated how natural selection, acting on small variations in polygenic traits, could push a population beyond the limit of either parent, and de Vries's mutation theory lost favor. Gradualism along with natural selection became firmly embedded in evolutionary theory, culminating with the Modern Synthesis of the 1940s. But, population genetics models changes occurring over relatively short periods of time compared to the amount of time represented by the fossil record. As George Gaylord Simpson wrote in his classic work *Tempo and Mode in Evolution*, population genetics "may reveal what happens to a hundred rats in the course of ten years under fixed and simple conditions, but not what happened to a billion rats in the course of ten million years under the fluctuating conditions of earth history."[14] Even with the success of the Modern Synthesis, it has always had critics. Geneticist Richard Goldschmidt and paleontologist Otto H. Schindewolf both argued that macroevolution could not be explained by microevolutionary mechanisms. Goldschmidt presented his views in the prestigious Silliman Lectures at Yale in 1939:

> Microevolution by accumulation of micromutations, is a process which leads to diversification strictly within the species, usually for the sake of adaptation of the species to specific conditions within the area which it is able to occupy. . . . [S]ubspecies . . . are not models for the origin of species. They are more or less diversified blind alleys within the species. The decisive step in evolution, the first step towards macroevolution, the step from one species to another, re-

quires another evolutionary method than that of sheer accumulation of micromutations.[15]

The reaction by the neo-Darwinians was immediate and harsh. Goldschmidt claimed that he "certainly had struck a hornet's nest. . . . This time I was not only crazy, but almost a criminal."[16] Ernst Mayr claimed that his own work on "geographical speciation and biological species was stimulated by opposition to Goldschmidt's proposed solution of speciation through systemic mutation."[17] Furious with Goldschmidt's 1940 book *The Material Basis of Evolution*, he claimed that "much of my first draft of *Systematics and the Origin of Species* was written in angry reaction to Goldschmidt's total neglect of such overwhelming and convincing evidence."[18] Mayr's opinion of Goldschmidt's work was shared by most evolutionary biologists. Although Mayr read Goldschmidt, many others did not. As Stephen Gould wrote, "Goldschmidt suffered the worst fate of all—to be ridiculed and unread."[19]

Gould was instrumental in getting Yale University Press to reprint *The Material Basis of Evolution*, and this served his own agenda of promoting Niles Eldredge's and his theory of punctuated equilibrium. Saltation has obtained its most significant hearing under the rubric of punctuated equilibrium. Indeed, Gould and Eldredge began their paper on punctuated equilibrium with Huxley's caution to Darwin on the eve of the publication of the *Origin*: "You have loaded yourself with an unnecessary difficulty in adopting *natura non facit saltum* so unreservedly."[20] A cynic might ask if anything had changed in evolutionary theory in 120 years.

Since Huxley's time a vast amount of knowledge has accumulated in support of evolution. Sampling of the fossil record is far more extensive and more carefully done. With the development of radiometric dating, it is possible to determine the age of fossils far more accurately. Yet in spite of the great advances in paleontology, explaining the pattern of the fossil record remains problematic. Although the details of the debate are different, three underlying issues have not changed since Huxley's time. First is the question of "gaps" in the fossil record. Today the fossil record is always cited in support of evolution, and while the record demonstrates unequivocally that the earth has been inhabited by vastly different organisms during different periods of earth history, it is not as clear as to how one group of organisms became replaced by another. During the synthesis period, Simpson identified as particularly difficult problems the origin of new "types" and structures; that is, the origin of higher taxa and the almost systematic absence of major "missing links." In spite of literally millions more fossils being found, gaps remain.[21] Gould and Eldredge maintain the Synthesis has not resolved this question: "Many breaks in the fossil record are real, they express the way in which evolution occurs, not the fragments of an imperfect record. The sharp breaks in a local column accurately records what happened in that area through time."[22] This problem may never be adequately resolved, since paleontologists estimate that less than 1 percent of all species end up as fossils. However, in spite of this incredibly incomplete record, we do continue to find transitional organisms such as *Basilosaurus*, a whalelike creature with feet. From the Eocene rocks in Pakistan we are accumulating a very good record of the transition in body form from a four-legged terrestrial ungulate to a fish-shaped whale.[23] Recently several remarkably well-preserved dinosaurs with feathers have been discovered in China, seeming to confirm Huxley's argument that dinosaurs were the connecting link between reptiles and birds. One looked so like *Archaeopteryx* that it was dubbed *Protoarcheopteryx*. *Caudipteryx* has long tail feathers, but its body looks more like *Velociraptor* than a chicken.[24] The "missing links" may indeed by rare, but they are not absent.

Related to, but distinct from the problem of gaps is the question of the tempo of evolution. The fossil record does not demonstrate a slow, gradual transition of one group to another.

The rate of evolution is jerky, resulting in a pattern that shows periods of rapid diversification interspersed with periods of relatively little change. By applying Mayr's theory of allopatric speciation to a geological time-frame, punctuated equilibrium challenges the Darwinian paradigm that most evolution occurs by gradual change within the phyletic mode. Instead, Eldredge and Gould argue that significant evolutionary change occurs primarily at the time of branching. Therefore, punctuated equilibrium is not a theory about mechanisms of change. Rather, it is a theory about rates of change that claims to give a better explanation for the particular pattern in the fossil record.

The question of both gaps and rates gives rise to the third question. If this pattern of fits and starts is accepted, does this represent a significant challenge to neo-Darwinian theory?[25] In spite of Gould asserting many times that punctuated equilibrium is not a theory about macromutation, nevertheless, it has reopened the discussion about macromutation. If the pattern is what Gould and Eldredge suggest, this implies that different mechanisms might be needed to adequately explain the pattern. Gould maintains that Goldschmidt's famous statement "The first bird hatched from a dinosaur's egg" is not quite as heretical as it sounds.[26] The systemic mutant may be incorrect, but Gould thinks the underlying idea is valid. Even small mutations, if they occur in regulatory genes or early in development, may have huge effects. Thus, Gould thinks that the actual genetic mechanism Goldschmidt suggested for his "hopeful monster" may have been incorrect, but the idea of a hopeful monster was not. Goldschmidt claimed that evolution resulted from inherited changes in development, and we now have evidence that for major phenomenon in evolution this is indeed the case. For instance, differences in Hox gene regulation appear to play a key role in the transition from fins to limbs.[27] Is something other than the principle of natural selection needed to explain the pattern of the fossil record? Contrary to what people such as Richard Dawkins and Daniel Dennett claim, an ever-increasing number of voices are answering un-equivocally *yes*!

No one disputes the power of natural selection to cause adaptation, that is, to shape organisms' form and function in order to maximize reproductive success. However, to claim that natural selection provides a complete explanation for the history of life and accounts for the diversity of forms seems limited at best. Can microevolutionary changes in gene frequency really be all that is needed to turn a reptile into a mammal or a fish into an amphibian? Furthermore, it is not at all apparent that either homology or diversity are fundamentally adaptive phenomena. The question of how form is generated has not been adequately answered. As many people have pointed out, embryology was not part of the Synthesis. Developmental biologists are calling for a second synthesis that will incorporate the findings of development and acknowledge the importance of a structural approach to solving basic biological problems.[28]

Structuralism has both strong and weak versions. The most extreme of the structuralists is Brian Goodwin, whose work in relationship to the morphogenetic field is discussed in the latter part of this chapter. Goodwin denies that the genetic program is what is primarily responsible for regulating development, claiming instead that gene products are important for stabilizing particular pathways of morphogenesis. In addition, Goodwin denies that selection is the driving force in the evolution of form. Rather, there are well-defined constraints that limit the kinds of transformations the fields can undergo, placing definite limits on the power of selection. Even among developmental biologists, Goodwin is considered by many an extremist. Rudolf Raff writes, "I see no support in the data for a strong version of structuralism."[29] I have chosen to focus on Goodwin's work because he has been the most outspoken and lays out starkly the kinds of problems that the functionalist approach has failed to adequately address.

Most developmental biologists support a weak version of

structuralism to explain much of the kind of constraint that is observed. In this version, first, there are certain generic processes that give rise to generic morphologies such as segments, and a gene-driven ontogeny eventually arises to stabilize and make such processes more precise. Second, there are only a limited amount of possible morphologies consistent with particular functions. For example, a tubular gut is probably the most efficient form for a digestive organ and, therefore, may have arisen several times independently. Furthermore, there are only so many ways that a tube can be generated from a sheet of cells. In summary, the weak version of structuralism claims that there is a finite number both of animal body plans and ontogenetic methods for achieving them. Early on during the Cambrian explosion, this range of forms was explored, and the same body plans have continued to be exploited over and over since there were few other possibilities left. While this version seems quite reasonable, Raff also sounds a word of caution about structuralism: "We can't be sure that we know the limits of morphology, nor that the forms of living animals extend to the limits of the possible." He cites as an example that because bacteria have high internal pressures they assume forms that have minimal surface area: spheres or rods with rounded ends. But in 1980 bacteria that looked like miniature postage stamps were discovered floating in the brine ponds of the Sinai desert. In Japan a related triangular bacterium was discovered living in the hot salt ponds in Japan. Like the two-dimensional creatures who inhabit the imaginary world of author Edwin Abott's *Flatland*, these organisms remind us that in diverse and harsh environmental conditions, it is difficult to know what kinds of forms are capable of emerging.[30] Just as in any thriving area of investigation, there needs to be a plurality of approaches in solving the problem of form as it relates to evolutionary theory. Regardless of the different points of view among developmental biologists, their work certainly needs to be examined to move evolutionary theory forward.[31]

It was not just that embryology was left out of the Synthesis. As Will Provine has written, it is more appropriate to refer to the evolutionary synthesis as the evolutionary constriction. Provine was discussing the constriction in a somewhat different context, arguing that the constriction resulted in evolutionary theory being purged of any purposive mechanisms (a development that Huxley would have applauded, unlike his grandson Julian), which made the conflict between evolution and religion inescapable.[32] The constriction cut down the variables that were relevant to the evolutionary process, all of which concerned the genetic modeling of populations. When in 1937 Theodosius Dobzhansky redefined evolution as changes in gene frequency, paleontology, along with every other biological discipline was relegated to mere description. Evolution became "the epiphenomenon of the genetics of populations."[33] Population genetics became the single discipline that could contribute anything meaningful in terms of theory. Today, a critical mass of biologists and philosophers of biologists are challenging this view, and evolutionary theory seems poised to undergo a radical revision.

Ironically, one call for revision is coming from the field of population genetics. The work of Japanese geneticist Motoo Kimura and American molecular biologists indicates that the vast majority of variation seen at the level of the DNA has no appreciable effect on phenotype.[34] In other words, the differences are not observable on the structural level. What could be more un-Darwinian than the neutral theory of evolution, which claims that most of the changes at the molecular level are not the result of Darwinian natural selection acting on advantageous mutants, but rather are due to the random fixation of selectively neutral or nearly neutral mutants through random genetic drift?

More relevant to the questions that interested Huxley, however, is another line of criticism that addresses many of the issues raised by the much maligned ideal morphologists of the eighteenth and nineteenth centuries. It is in this context that Huxley's "pre-Darwinian" ideas become timely. Key to the reevaluation of modern evolutionary theory is (1) the study of complex systems, (2) a reexamination of the meaning and evidence of homology, and (3) a resurgence of the idea of the morphogenetic field (which will be discussed shortly).

The Science of Complexity

Stuart Kauffman and his associates at the Santa Fe Institute have been pioneers in the emerging discipline that studies complex systems. Like Huxley, Kauffman does not deny the importance of natural selection, but he argues that natural selection "has not labored alone to craft the fine architecture of the biosphere, from cell to organism to ecosystem."[35] Kauffman has come to believe that the root source of the order we see in the biological world comes from self-organization. It arises naturally and spontaneously. Rather than natural selection tinkering with random variations to craft an organism, to "build a better mouse trap," or in this case a mouse, natural selection chooses among certain forms, forms that were generated by laws of complexity. Kauffman and his colleagues are trying to discover these laws of self-organization. In this view organisms are not just "tinkered together contraptions, but expressions of deeper natural laws."[36] It is tempting to say that Kauffman is searching for Spencer's "law of evolution," however, this is not quite right. Spencer wanted a theory of evolution in the Darwinian sense, a theory of *change*. Kauffman certainly considers himself an evolutionist, but rather than a theory of change Kauffman is searching for a theory of structure. His "laws of self-organization" may provide boundary conditions or constraints that shape the kinds of change that may occur, but he is not trying to describe a theory that will explain what sort of changes or trends are expected to occur. Spencer, however, did think that the deeper law of evolution would explain specific trends, for example, from the homogenous to the heterogeneous, from the simple to the complex. If these laws of emergent order exist, are they compatible with the random mutations and selection of Darwinism, which implies that life is contingent, accidental, and unpredictable? How did such complexity in the universe emerge from the Big Bang 15 billion years ago? Scientists are familiar with two major forms by which order arises. The first involves low-energy equilibrium systems such as a ball that rolls to the bottom of a bowl, wobbles, and stops. The kinetic energy of motion acquired because of gravity is dissipated into heat by friction, and the ball stops at a position that minimizes its potential energy. The ball is at equilibrium, and no further input of energy is needed to maintain its position at the bottom of the bowl.

The second way order arises requires a constant source of mass or energy or both to maintain the ordered structures. Such a system is a nonequilibrium system. An example is the whirl in the drain of a bathtub, which can be stable for long periods of time if water is continuously added to the tub and the drain is left open. A quite dramatic example of a nonequilibrium structure is the Great Red Spot vortex on Jupiter, which is essentially a storm system that has existed for several centuries. It is a stable structure of matter and energy through which both matter and energy flow. There are intriguing similarities between the Red Spot and a living organism, since the individual molecular constituents in both change many times while still maintaining an overall stable structure. Although the behavior of nonequilibrium chemical systems have been well studied, their interpretations remain controversial. Since organisms are nonequilibrium systems, discovering general laws that would predict the behavior of such systems would provide deep insight into understanding the living world.[37]

The Great Red Spot is just one example of a chemical nonequilibrium system that shows similarities to living systems. Another example is the concentric circular and spiral wave patterns generated by the famous Beloussov-Zhabotinsky reaction that are amazingly similar to the patterns generated by the aggregation of individual amoebas in the life cycle of the cellular slime mold. The slime mold has several distinct stages in its most unusual life cycle. At one stage it consists of single, swimming, flagellated swarm cells that will lose their flagella and change into an amoeboid form. Thousands of these amoebalike cells merge, lose their cell boundaries, and form a multinucleate mass called a plasmodium, which eventually turns into a hard, dry, fruiting body containing spore cases, which eventually burst. The spores germinate into swarm cells, completing the cycle. The aggregation of the amoeboid cells is caused by a chemical signal, the release of cAMP. In one case the pattern is caused by a purely chemical process, while the other arises from cell interactions.[38] Are these basically similar processes or just an amazing coincidence? Similar patterns are generated in the behavior of ant colonies, brains, and hearts.[39] Certainly in terms of molecular composition, the Beloussov-Zhabotinsky reaction and these other systems have nothing in common. The key to this unexpected emergent order appears to be in the relationship of the molecules to each other, rather than their composition.

This kind of research suggests that the spontaneous order that arises in complex physical systems can account for much of the order found in organisms. If this is indeed correct, it means that the role of natural selection in shaping the pattern of evolution is more limited than what has generally been assumed. What is selected in biological evolution reflects a compromise between selection and the spontaneous properties of the class of systems that selection is acting on. Selection is a "combinatorial optimization process in a rugged fitness landscape with many peaks, ridges, and valleys." The typical structure of the landscape along with population flow upon the landscape under the drives of mutation and selection means that achieving and maintaining rare, highly adaptive states usually does not occur. Thus, many of the features in organisms are a reflection of the generic properties of an entire class of systems under selection rather than the particular success of selection. In other words, "the spontaneous properties of complex systems under selection will often be similar to those generic in the class of systems under selection, not *because* of selection, but *despite* it." This in turn suggests that an analysis of these generic properties can be expected to be predictive of features of organisms, without knowing the specific details.[40]

What kind of data does Kauffman have to support his ideas? To try and discover these laws of complexity, he has been running powerful computer simulations that model genetic regulatory networks, and he has concluded that the genetic programs underlying ontogeny have powerful "self-organized structural and dynamical properties." His results suggest that the order generated in these self-organizing systems prevents natural selection from moving them very far from their generic state. Selection can choose only from certain limited alternatives that are determined by the dynamics of that system. In other words, selection is not a creative force. Although Kauffman is only running computer simulations, his models have been able to make predictions that are surprisingly accurate. For example, his models predict that the number of cell types an organism has increases as a square root function of the number of genes of that organism. This is what is actually observed across many phyla. His models also imply that any cell type can flow only to a few neighboring cell types and then on to a few other cell types. Such a result suggests that ontogeny should occur along branching developmental pathways, and indeed, all contemporary multicellular organisms develop along just such pathways.[41] Furthermore, we are now accumulating increasing amounts of evidence that indicate that basic developmental pathways have been conserved among phyla that have been independently evolving for at least 500 million years. For instance, the pathway for eyes is controlled by the same genes in flies, squid, and vertebrates. Another dramatic finding has been the dis-

covery of the "homeobox" gene complex, which is responsible
for creating the anterior-posterior axis in development in
insects. However, it was soon discovered that fruit flies, frogs,
mice, and humans as well as other vertebrates all have the
homeotic gene complex arranged in the same order on their
chromosomes and that the anterior-posterior expression pattern of the individual genes was the same in flies and vertebrates. Deleting one or more of these genes could produce
atavistic changes such as the formation of reptilian jaw and neck
vertebrae in mice. Although insects and vertebrates create their
body axes, limbs, and nervous systems in quite different ways,
homologous gene complexes appear to be controlling the
development of body plans in a similar way in virtually every
single animal on the planet.[42] Not quite two hundred years ago
Geoffroy St.-Hilaire claimed that all animals were variations on
a single basic body plan. Today, the unity of type is finding support in the detailed studies of molecular genetics.

A theoretical debate rages as to the meaning of these findings. Is the conservation of developmental pathways the result
of selection, or is it a property of generic systems that, in fact,
is impervious to selection? Strict adaptationists argue that such
conservation occurs because the optimality of these pathways
has been reached and can't be improved upon. But Kauffman
believes that such organization is a "property which is so deeply
embedded in the entire ensemble of genomic regulatory systems accessible to selection, so deep a property of parallel pro-
cessing nonlinear dynamical systems, that selection cannot
avoid this property."[43] Many developmental biologists concur
with such an analysis. Developmental pathways limit what is
possible. Not just any form can evolve, no matter how adaptive
it might be. Even before this molecular evidence was available,
many paleontologists had been sympathetic to such a view.
They borrowed from anatomists the idea of the *Bauplan*. The
Bauplan, they argue, suggests that developmental constraints
limit the extent to which organs can undergo functional adap-

tation. Furthermore, the origin of a new *Bauplan* requires the reorganization of developmental systems. The concept of the *Bauplan* is useful, writes Jeffrey S. Levinton, "because of the observation that evolution has involved elaborations on a relatively small number of combinations of major body parts, rather than of all possible combinations of characters."[44] Paleontologists refer to the possible morphologies that organisms can occupy as "morphospace." The pattern of the fossil record documents that not all theoretically available morphospace is filled. A variety of answers have been offered to explain the so-called gappiness of the fossil record. Functionalists argue that the missing forms are selected against. Historicists claim that a particular lineage is constrained historically and hasn't diffused sufficiently to fill the space. Contingency and chance play a large role in their interpretation of the fossil record. However, a complete explanation for the pattern of the fossil record undoubtedly needs to draw on the research of developmental biologists as well. A crucial aspect of this research has been the reassessment of the meaning of homology.

The Meaning of Homology

It is indeed ironic that the importance of homology was eclipsed with the rise of population genetics, as it was always a central idea to Darwin. For him embryology provided the strongest evidence in favor of evolution. In a letter to Huxley he asserted,

> I rather doubt whether you seen how far, as it seems to me, the argument for homology and embryology may be carried. I do not look at this as mere analogy. I would as soon believe that fossil shells were mere mockeries of real shells as that the same bones in the foot of a dog and wing of a bat, or the similar embryo of mammal and bird, had not a direct significa-

tion, and that the signification be unity of descent or nothing.[45]

I rather doubt if Huxley underestimated the importance of homology. As was pointed out in chapter 2, Huxley believed that providing a rigorous definition of the meaning and distinction between homology and analogy would be the key to discovering the general laws that underlie the generation of animal form. There is no shortage of modern biologists who concur with such a view. As Colin Patterson has written, "All useful comparisons in biology depend on the relation of homology."[46] Brian Hall has recently edited a large volume on the meaning of homology.[47] David Wake claims that whatever homology means,

> it is the central concept for *all* biology. Whenever we say that a mammalian hormone is the "same" as a fish hormone, that a human sequence is the "same" as a sequence in a chimp or a mouse . . . even when we argue that discoveries about a roundworm, a fruit fly, a frog, a mouse or a chimp have a relevance to the human condition—we have made a bold and direct statement about homology.[48]

However, as Wake implies, the meaning of homology is not unproblematic. Huxley had already began to grapple with some of the difficulties in defining homology rigorously. Pre-Darwin, drawing on the ideas of Geoffrey and particularly Richard Owen, homology was generally regarded as a morphological correspondence that was determined by relative position and connection. However, Huxley asked, what if animals agreed in structure but differed in development, or the reverse—agreed in development but differed in the final structure? Which should take precedence? Huxley decided that development was the key to deciding whether a structure should be regarded as homologous or merely analogous. Although Darwin retained

Owen's definition of homology and drew heavily on embryology to advance his evolutionary argument, his explanation of homology was quite different. The "essential" similarities that were observed in different organisms were the result of having a common ancestor. Then an interesting and not uncommon transition took place—from explanation to definition. While at first common ancestry was the explanation for homology and thus homology could be used as evidence for evolution and a phylogenetic relationship, it then became part of the definition. Homology was *defined* as any similarity that could be traced to a common ancestor. However, as Huxley pointed out in his discussion of classification, certain problems arise once this connection is made between homology and evolution. Often similar structures arise independently. In contrast, very different structures often arise through evolutionary transformation. In determining genealogies, one has to rely in large part on similarity of characters. Thus, one has to make a choice as to which criterion should be given primacy—similarity or ancestry. In addition, homology was now being used to explain both the maintenance of similarity *and* the transformation of form.

Sir Galvin de Beer further elaborated the problems that occur when homology is linked to ancestry in its definition. As he pointed out, the similarities of form between the fore and hind limbs of any species "is not real homology, as fore-limb and hind-limb cannot be traced back to any ancestor with a single pair of limbs."[49] Rather, fore limbs and hind limbs appear to have evolved independently from the pectoral and pelvic fins in a primitive gnasthostome. However, this would lead to the unsatisfactory conclusion that human arms, bats' wings, and the fore limbs of a horse are homologous, but human arms and legs are not.

De Beer suggested that perhaps there was a genetic basis for homology. Modern research has certainly confirmed this but has further complicated the picture. Not only do we see similar genes acting in lineages that have been evolving independently for hundreds of millions of years, but because most genes are pleiotropic in their action, a mutation in a single gene often causes morphological changes in structures that clearly are *not* homologous. In addition, identical morphology of particular characters of different individuals does not necessarily mean that the underlying genotype is identical. For example, the *eye-less* mutation in *Drosophila* results in the eyes failing to form. However, after several generations, normal eye formation returns due to changes in other genes, even though the *eyeless* mutation is still present.

Finally, the results from developmental genetics have created complete havoc with our traditional distinction between analogy and homology. The segmentation in the body parts of flies and vertebrates has always been cited as a classic example of analogy. However, with the discovery of the homeobox genes we now have "homologous" genes responsible for "analogous" processes. The vertebrate eye and the insect eye were also regarded as analogous structures. However, they are both based on the expression of the *Pax*-6 gene. It is believed that along with cephalopod eyes, they are all descendants of a basic metazoan photoreceptive cell that was regulated by the *Pax*-6 gene. One of the most dramatic examples of homology of process is the receptor tyrosine kinase-ras signal transduction pathway that has been found in mice, nematodes, and fruit flies. The pathway is so similar that many of the components can be interchanged between species, but it is involved in generating entirely different structures. In the fruit fly, it is part of the developmental pathway for both the retina and the terminal body segments. However, the same pathway is involved in the determination of the nematode vulva and the mammalian epidermis as well.[50]

> Whereas classic homology has been one of structure—be it of skeletons or genes—the homology of process goes into the very mechanism of development. Whereas classical homology

looks at the similarities between entities, the homology of process concerns the similarities of dynamic interactions.[51]

Thus, organs such as the vertebrate and arthropod eye may be structurally analogous, but they have been formed by homologous processes.

The Return of the Morphogenetic Field

Homology is about what is conserved in evolution. How and why particular features are conserved, however, is a problem distinct from identifying homologues. The discovery of homologies of process has brought a renewed interest in trying to answer these questions and has contributed to the resurgence of another important idea in embryology: the morphogenetic field. The concept of fields was made popular at the end of the nineteenth and first part of the twentieth centuries with the work of embryologists Hans Driesch, Theodor Bovari, and others, but fell into disrepute in part because of its vagueness and also the somewhat mystical connotations with which it was associated. As a descriptive term, however, the idea of the morphogenetic field was quite useful. Fields were areas of tissue in the embryo that did not seem to be differentiated, yet were destined to give rise to specific structures in the course of development such as limbs, heart, lens. Experiments with fields produced quite interesting results. If divided, they would give rise to duplicate structures, and if damaged they could regenerate. If they were transplanted to another site or animal, fields would produce their own characteristic structure in the host. Recent work has shown that these properties are consistent with the modes of action and expression of major regulatory genes. For example, limb fields on the flanks of the *Drosophila* embryo are first recognized by the expression of the gene *Distal-less*, which is essential for the development of appendages.

It is worth pointing out that processes that are being defined as homologous are interactions between groups of cells that in the older literature had been referred to as fields such as the limb field, the eye field, etc. For biologists who have retained this terminology, usually all they meant was that these are areas of tissue that are destined to form those particular structures. However, Brian Goodwin uses the concept in a way that is much more similar to the older concept as described by Wilheim Roux and Hans Driesch. Goodwin's research program emphasizes the influences that affect the spatial order that emerges from cell-cell interactions. He, like Kauffman, argues that the order we see in living nature is an expression of properties intrinsic to complex dynamic systems. Fields are generated from simple rules of interaction among a variety of elements. Rather than using genealogy to define homology, Goodwin claims that the experimental study of the processes of development will yield generative principles that show homology to be the result of developmental dynamics "that has the logical structure of an equivalence relationship as used in mathematics."[52]

Ever since Darwin, history has been intrinsic to our definition of homology, and thus Goodwin's repudiation of history in trying to discover the laws of form is quite radical. In a somewhat polemical example, he claims that if we answered the question "Why does the earth go around the sun in an elliptical orbit?" by stating that it did last year and the year before, all the way back to the beginning of the solar system, and nothing exists to change it, most people would not find that explanation satisfactory. However, this is a very common type of explanation in biology. Goodwin maintains that if a true integration of development and evolution is to occur, the actual generative processes that underlie stable life cycles must be understood. Combining the results of such a research program with that of reconstructing phylogenies, we will then be able to answer the question whether the relationship of taxa to one another is primarily the result of history, or if instead it is the result of generic states of ontogenetic dynamics and, thus, is largely independent of history.[53] Nevertheless, it will be very difficult to determine whether the stability of the shared dynamics is the result of developmental inertia or of selection. Goodwin is searching for a theory of morphogenesis that will provide the same explanatory power as Newton's laws, which united celestial and terrestrial objects into a vast mechanical system—all of which moved in perfect harmony and the connections could be expressed in mathematical terms. This is an extremely ambitious agenda, and Goodwin admits that it is not clear whether such a program can succeed. It is a program that has its roots in pre-Darwinian morphology and addresses issues that many biologists, from Huxley to Dreisch to de Beer, believe have not been adequately explained by natural selection. However, it is not a program that is antagonistic to Darwinian natural selection, but rather is complementary. For Huxley, the most important goal in zoology was to define the doctrine of animal form. Goodwin, like Huxley, believes that such a goal will be realized by discovering the laws or rules that generate form.

Goodwin has devoted a significant part of his research career to studying the life cycle of the alga *Acetabularia acetabulum*, a member of the group *Dasycladales*. Characteristic of the alga and its ancestors is the development of structures known as whorls. However, the alga does not seem to need them and sheds them soon after it makes them. Why does it not get rid of them altogether? The standard answer is that it is a trait inherited from the ancestor and that making whorls is just too deep and persistent a property of morphogenesis to be gotten rid of. Thus, whorls, although not needed by the species who make them, are carried along by a kind of developmental inertia. Goodwin does not deny that this is a plausible explanation but instead claims that this points to a question rather than answers it. Why are whorls made in the first place, and why are they so difficult to get rid of? Natural selection tells us that they are useful for most members of the order as gametophores, they are used for photosynthesis, and since they do not "cost" too much to make, they are maintained. However, this explanation in terms of natural selection and history merely reiterates what is observed in terms of function and costs. According to Goodwin, the answer will not be found in history but in discovering the laws that generate these structures.

Without going through the details of how *acetabularia* makes whorls, Goodwin, working with several colleagues, most notably Lynn Trainor and Christian Brière, has developed equations to describe the morphogenetic field responsible for making whorls. They have drawn on the work of many others who have developed mathematical descriptions of morphogenesis, have investigated the properties of the cytoskeleton, particularly as it relates to calcium gradients, and have examined the growth of plant cell walls. These results showed that starting from spatially uniform states, stable patterns of calcium gradients and mechanical states emerged spontaneously as a result of the interactions among various factors. From initially simple forms complex morphology emerged.[54] Notice that this emergence of form did not involve genes. Computer simulations have allowed them to explore the types of form that a cell such as a developing alga could produce. This research indicates that morphogenetic fields, in spite of their great diversity, particularly in the varied adult morphologies they produce, are, nevertheless, "limited in their possible transformations by intrinsic organizational constraints."[55]

These investigations suggest that the taxonomies that we have developed to describe the relationship of organisms to one another are not the result of trial-and-error tinkering by natural selection, but rather reflect a deep pattern of ordered relationships. Although a great deal of variety exists in the shape of whorls between species and in a single whorl as the individual alga matures, they still represent variations on a theme. There is an inherent stability of this basic form, which is a persistent structure of the group and is what makes possible a systematic classification of the group. It is the actual process of growth involving a series of feedback loops that is a major stabilizer of the pattern that is seen. The nonlinear dynamics of the interaction of parts along with the change of shape results in a robust sequence of events that generates a general morphological pattern that gives the entire group of *Dasycladales* its taxonomic unity. From this perspective, then, *Dasycladales* are a natural group not because of their history, but rather because of the way their basic structure is generated. This does not mean that history doesn't matter. The historical sequence of how individual species evolved can give us insight into the role that genes play in generating particular forms. Nevertheless, this analysis implies that a morphogenetic theory that describes how different forms are generated is the key to answering the question that paleontologists have posed for some time: Why is much of the available morphospace not filled? The answer is not one that is grounded in natural selection and the contingencies of history, but it rather claims that only certain morphologies are possible because development proceeds according to certain well-defined principles that set the boundaries of morphogenetic fields.

Goodwin's approach minimizes the role of both history and gene action in his concept of fields. It is a program to counter the overwhelming dominance of the view that everything can be explained in terms of DNA. However, developmental biologists, including Goodwin, who are calling for a reevaluation of the Synthesis certainly recognize that genes and history do matter in the phenotype that ultimately is produced. The stable life cycle that is repeated from generation to generation depends on the inheritance of DNA, the developmental system, and the environment. Furthermore, the whole reformulation of the meaning of homology is based on the results in developmental genetics. In Huxley's time homology was defined in terms of similarity of organ and skeletal systems. Today, we find homology existing at a very deep level—at the level of the gene.

The homologies of process that exist in morphogenetic fields, in fact, are the best evidence we have for evolution. However, such results also suggest that the research program in population genetics must be reformulated to look at the role of regulatory genes if population genetics is going to remain relevant to evolutionary theory.

In this research program, which is attempting to fully integrate development with evolution, it is the morphogenetic field that emerges as the primary unit of both ontogenetic and phylogenetic change. Fields are discrete units of embryonic development that result from the dynamic interactions of a variety of elements, including genes and gene products. Fields can be limited by diffusion gradients, cell adhesion molecules, and a variety of other factors. Changes in any of these parameters can result in different phenotypes and evolutionary novelty. The field is what is primarily responsible for generating the complex anatomical structures we see in organisms. Although genes play a significant role in specifying the nature of a particular field, fields also have properties that are independent of the gene content. In addition, homologous genes can play different roles in different fields, and thus fields act like an ecosystem in which the genes function. Claiming that the morphogenetic field is a major element of both developmental and evolutionary change shifts the emphasis away from a solely gene-oriented model of evolution and development. In such a model natural selection plays a much smaller role in evolution than has been believed. It is merely the filter for unsuccessful morphologies that are generated by development.[56]

Conclusion

E. S. Russell, in his classic work *On Growth and Form*, documented that one of the most fundamental debates in nineteenth-century biology was over what is primary: form or function.

Huxley, trained in the tradition of developmental morphology, but impressed with the explanatory power of Darwin's theory, recognized that each program had much to offer. With the rise of population genetics vindicating the principle of natural selection, the functional approach has dominated the evolutionary research program of the twentieth century. However, recent research in developmental biology returns to many of the issues surrounding the form/function debates of the nineteenth century. It also suggests that perhaps the theoretical debate over whether form determines function or vice versa is misguided. Homology of process demonstrates that the two are intimately interconnected at the very deepest level. Morphology is the product of the interdependent relationship between form and function; the position and arrangement of organs are the expression of functional laws of organisms. Putting the theoretical question aside, as a practical matter, Huxley believed that studying how form comes to be generated, that is, by studying development, was a better research strategy to understand function and adaptation than the other way around. We do not need to make a judgment whether it is a better approach. Clearly the functional approach has been a powerful one, but it is also apparent that it still has not adequately answered profound questions relating to the nature of form.

One of the great paradoxes of evolution is that with the great diversity of forms that have evolved in the last half billion years, there also has been tremendous stability. Body plans that exist today are not easily transformable to one another, and all the animal forms that exist are variants on thirty-five basic body plans that appeared in the Cambrian. Huxley believed the key to these questions lay in development. Today, more and more people share his view. Goodwin and Webster, in fact, argue that evolution provides only limited insight into the problem of form. They believe that a causal explanatory theory of form and the relationships of forms to one another, a theory of morphogenesis, will not be supplemental, but will be as fundamental to biology as the theory of evolution. Huxley, I'm sure, would agree.

Notes

1. Marjorie Grene, *The Understanding of Nature, Essays in the Philosophy of Biology* (Dordrecht: Reidel, 1974), p. 408. Quoted in Gerry Webster and Brian Goodwin, *Form and Transformation* (Cambridge: Cambridge University Press, 1996), p. 3.

2. Webster and Goodwin, *Form and Function*, p. 129.

3. Michael Bartholomew, in "Huxley's Defense of Darwinism," *Annals of Science* 32 (1975): 335.

4. Adrian Desmond's recent biography, *Huxley* (New York: Addison Wesley, 1997), acknowledges that Huxley used morphology to further the cause of evolution.

5. P. F. A. Maderson et al., "The Role of Development in Macroevolutionary Change," in *Evolution and Development*, John T. Bonner, ed. (New York: Springer-Verlag, 1982), p. 180.

6. Brian Goodwin, *How the Leopard Changed Its Spots* (New York: Charles Schribner's Sons, 1994), p. viii.

7. Ibid.

8. Thomas Huxley, "Evolution in Biology," in *Science and Culture* (1878; New York: D. Appleton & Co., 1882), p. 289.

9. Ibid., p. 298.

10. Ibid., p. 304.

11. See Daniel Brooks and E. O. Wiley, *Evolution as Entropy* (Chicago: University of Chicago Press, 1986); David Depew and Bruce Weber, *Darwinism Evolving* (Cambridge, Mass.: MIT Press, 1995); Stanley Salthe, *Development in Evolution: Complexity and Change in Biology* (Cambridge, Mass.: MIT Press, 1993); and Rod Swenson, "Thermodynamics and Evolution," in *The Encyclopedia of Comparative Psychology*, G. Greenberg and M. Haraway, eds. (New York: Garland Publishing, 1997).

12. Stuart Kauffman, "Origins of Order in Evolution, Self-Organisation, and Selection," in *Theoretical Biology*, Brian Goodwin and Peter Saunders, eds. (Edinburgh: Edinburgh University Press, 1989), p. 67.

13. Huxley, "Evolution in Biology," p. 314.

14. George Gaylord Simpson, *Tempo and Mode in Evolution* (New York: Columbia Press, 1944), p. xvii.

15. Richard Goldschmidt, *The Material Basis of Evolution* (New Haven: Yale University Press, 1982), p. 183.

16. Goldschmidt, *Material Basis of Evolution*, quoted by Gould in his introduction to *The Material Basis of Evolution*, p. xiv.

17. Ernst Mayr, *The Growth of Biological Thought* (Cambridge, Mass.: Belknap Press, 1982), p. 381.

18. Ernst Mayr, "How I Became a Darwinian," in *The Evolutionary Synthesis*, Ernst Mayr and William Provine, eds. (Cambridge, Mass.: Harvard University Press, 1980), p. 421.

19. Goldschmidt, introduction to *The Material Basis of Evolution*, p. xiv.

20. Thomas Huxley, November 23, 1859, *Life and Letters of T. H. Huxley* (*LLTHH*), 2 vols., Leonard Huxley, ed. (New York: D. Appleton & Co., 1900), quoted in Stephen Gould and Niles Eldredge, "Punctuated Equilibria: The Tempo and Mode of Evolution Reconsidered," *Paleobiology* 3 (1977): 115–21.

21. George Gaylord Simpson, "Biographical Essay," in Mayr and Provine, *Evolutionary Synthesis*, p. 457.

22. Niles Eldredge and Stephen Gould, "Punctuated Equilibria: An Alternative to Gradualism," in *Models in Paleobiology*, Thomas Schopf, ed. (San Francisco: Freeman, Cooper and Co., 1972), p. 96.

23. See Rudolf Raff, *The Shape of Life* (Chicago: University of Chicago Press, 1996), pp. 400–404.

24. See J. Fischman, "Feathers Don't Make the Bird," *Discover* (January 1999): 48–49.

25. Even Brian Charlesworth et al., in a harsh attack on punctuated equilibria, do not dispute the punctuational view of the fossil record, but rather if something other than neo-Darwinian theory is needed to explain the pattern. See Brian Charlesworth, Russell Lande, and Montgomery Slatkin, "A Neo-Darwinian Commentary on Macroevolution," *Evolution* 36 (1982): 474–98.

26. Gould, *The Material Basis of Evolution*, p. xxxvii.

27. Scott Gilbert, "Conceptual Breakthroughs in Developmental Biology," *Journal of Biological Science* 23, no. 3 (1998): 169–76.

28. Scott Gilbert, John Opitz, and Rudolf A. Raff, "Resynthesizing Evolutionary and Developmental Biology," *Developmental Biology* 173 (1996): 357–72.

29. Raff, *The Shape of Life*, p. 312.

30. Ibid, pp. 314–15.

31. There is an enormous literature in developmental biology. In addi-

tion to Raff's book and the other citings in this chapter, a few relevant citings on the relationship of development to evolution include Pere Alberch, "Developmental Constraints in Evolutionary Processes," in *Evolution and Development*, John T. Bonner, ed. (Berlin: Springer-Verlag, 1982), pp. 313–32; Wallace Arthur, *A Theory of the Evolution of Development* (Chichester: John Wiley and Sons, 1988), and *The Origin of Animal Body Plans* (Cambridge: Cambridge University Press, 1997); R. A. Raff and E. C. Raff, eds., *Development as an Evolutionary Process* (New York: Alan Liss, 1987); and A. Seilacher, "Early Multicellular Life, Late Proterozoic Fossils and the Cambrian Explosion," in *Early Life on Earth*, S. Bengston, ed. (New York: Columbia University Press, 1994), pp. 389–400.

32. William Provine, "Progress in Evolution and Meaning in Life," in *Evolutionary Progress*, Matthew Nitecki, ed. (Chicago: University of Chicago Press, 1989), pp. 49–74. See also Marc Swetlitz, "Julian Huxley and the End of Evolution," *Journal of the History of Biology* 28 (1995): 181–217.

33. Gilbert, Opitz, and Raff, "Resynthesizing," p. 358.

34. Motoo Kimura, *The Neutral Theory of Molecular Evolution* (Cambridge: Cambridge University Press, 1983).

35. Stuart Kauffman, *At Home in the Universe* (New York: Oxford University Press, 1995), p. vii.

36. Ibid, p. 8.

37. Ibid., pp. 20–24.

38. Goodwin, *How the Leopard Changed Its Spots*, pp. 45–52. See also G. Nicolis and I. Prigone, *An Introduction to Complexity* (New York: W. H. Freeman, 1987).

39. O. Miramontes, R. Sole, and B. Goodwin, "Collective Behavior of Random-Activated Mobile Cellular Automata," *Physica* 63 (1993): 145–60; C. A. Skarda and W. F. Freeman, "How Brains Make Chaos in Order to Make Sense of the World," *Behavior and Brain Science* 10 (1989): 161–95. Goodwin, *How the Leopard Changed Its Spots*, pp. 63–64. See also A. T. Winfree, *The Geometry of Biological Time* (New York: Springer-Verlag, 1989).

40. Kauffman, "Origins of Order," pp. 68–69.

41. Ibid., pp. 79–80.

42. Gilbert, Opitz, and Raff, "Resynthesizing," pp. 363–64.

43. Kauffman, "Origins of Order," p. 84.

44. Jeffrey S. Levinton et al., "Organismic Evolution: The Interaction of Microevolutionary and Macroevolutionary Processes," in *Patterns and Processes in the History of Life*, David Raup and David Jablonski, eds. (Berlin:

Springer-Verlag, 1986), p. 168. In the same volume see also J. Valentine, "Fossil Record of the Origin of Bauplan and Its Implications," pp. 209–31. Ernst Mayr also acknowledges the importance of the concept of *bauplan* in recent controversies over macroevolution. See Ernst Mayr, *Towards a New Philosophy in Biology* (Cambridge, Mass.: Harvard University Press, 1988), pp. 406–407.

45. Charles Darwin, quoted in Dov Ospovot, *The Development of Darwin's Theory* (New York: Cambridge University Press, 1981), p. 165.

46. Colin Patterson, *Molecules and Morphology in Evolution* (Cambridge: Cambridge University press, 1987), p. 18, quoted in M. Donoghue, "Homology," in *Key Words in Evolutionary Biology*, E. Fox-Keller and E. Lloyd, eds. (Cambridge, Mass.: Harvard University Press, 1992), p. 170. See, however, the entire entry for a good treatment of the changing definition of the word "homology."

47. Brian K. Hall, *Homology* (New York: Academic Press, 1994).

48. David Wake, "Comparative Terminology," *Science* 265 (1994): 268–69. Quoted in Gilbert, Opitz, and Raff, "Resynthesizing," p. 362.

49. G. de Beer, *Homology an Unsolved Problem* (Oxford: Oxford University Press, 1971). Quoted in Webster and Goodwin, *Form and Function*, pp. 140–41.

50. For further examples and a more detailed discussion of the homology of process, see Gilbert, Opitz, and Raff, "Resynthesizing," pp. 364–65.

51. Ibid., p. 364.

52. Webster and Goodwin, *Form and Transformation*, p. x.

53. Goodwin and Saunders, *Theoretical Biology*, p. 98.

54. Goodwin, *How the Leopard Changed Its Spots*, chapter 4.

55. Webster and Goodwin, *Form and Transformation*, p. 344.

56. Gilbert, Opitz, and Raff, "Resynthesizing," p. 368.

Bibliography

Manuscript Collections

American Philosophical Society, Philadelphia   Thomas Huxley
British Museum of Natural History, London   Richard Owen
Edinburgh Library, Special Collections, Edinburgh   Charles Lyell
Imperial College of Science and Technology, London   Thomas Huxley
Linnean Society, London   Thomas Huxley
    William MacLeay
Royal Botanical Gardens, Kew   Joseph Dalton Hooker
Royal College of Surgeons, London   Richard Owen
University Library, Cambridge   Charles Darwin

Books and Articles

Agassiz, Louis. "Evolution and Permanence of Type." *Atlantic Monthly* 33 (1874): 92–101.

Ainsworth Davis, J. R. *Thomas H. Huxley.* New York: E. P. Dutton & Co., 1907.

Alberch, Pere. "Developmental Constraints in Evolutionary Processes." In *Evolution and Development,* John Tyler Bonner, ed., pp. 313–32. Berlin: Springer-Verlag, 1982.

Albritton, G. G., ed. *The Fabric of Geology.* Stanford, Calif.: Freeman, Cooper, and Co., 1963.

Allen, Colin, Marc Bekoff, and George Lauder, eds. *Nature's Purposes.* Cambridge, Mass.: MIT Press, 1998.

Amundson, Ron. "Typology Reconsidered: Two Doctrines on the History of Evolutionary Biology." *Biology and Philosophy* 13 (1998): 153–77.

Appel, Toby. *The Cuvier-Geoffrey Debate*. New York: Oxford University Press, 1987.

Arthur, Wallace. *A Theory of the Evolution of Development*. Chichester: John Wiley and Sons, 1988.

———. *The Origin of Animal Body Plans*. Cambridge: Cambridge University Press, 1997.

Ashforth, Albert. *Thomas Henry Huxley*. New York: Twayne, 1969.

Asma, Stephen. *Following Form and Function*. Evanston, Ill.: Northwestern University Press, 1996.

Ayala, Francisco, ed. *Molecular Evolution*. Sunderlan, Mass.: Sinauer Associates, 1976.

Baer, Karl Ernst von. "Philosophical Fragments *Über Entwickelungsgeschichte The Fifth Scholium*," 1828. *Scientific Memoirs, Natural History*, A. Henfrey and T. Huxley, eds. London: Taylor and Francis, 1853.

Barnett, Samuel A., ed. *A Century of Darwin*. London: William Heinemann LTB., 1958.

Barr, Alan, ed. *Thomas Henry Huxley's Place in Science and Letters: Centenary Essays*, 2 vols. Athens: University of Georgia Press, 1997.

Barrett, Paul, Peter Gautrey, Sandra Herbert, David Kohn, and Sydney Smith, eds. *Charles Darwin's Notebooks, 1836–44*. Ithaca, N.Y.: Cornell University Press, 1987.

Barry, Martin. "On the Unity of Structure in the Animal Kingdom," *Edinburgh New Philosophical Journal* 22 (1837): 116–41.

Barthlomew, Michael. "Lyell and Evolution: An Account of Lyell's Response to the Prospect of an Evolutionary Ancestry for Man." *British Journal for the History of Science* 6 (1973): 284.

———. "Huxley's Defense of Darwin." *Annals of Science* 32 (1975): 525–35.

———. "The Non-Progress of Non Progression: Two Resonses to Lyell's Doctrine." *British Journal for the History of Science* (1976): 166–74.

———. "The Singularity of Lyell." *History of Science* 17 (1979): 276–93.

Barton, N. H., and B. Charlesworth. "Genetic Revolutions, Founder Effects, and Speciation" *Annual Review of Ecology and Systematics* 15 (1984): 133–64.

Barton, Ruth. "Evolution: The Whitworth Gun in Huxley's War for the Liberation of Science from Theology." In *The Wider Domain of Evolutionary Thought*, D. Oldroyd and I. Langham, eds., pp. 261–87. The Netherlands: D. Reidel Publishing Co., 1983.

Bateson, William. *Materials for the Study of Variation: Treated with Especial*

Regard to Discontinuity in the Origin of Species. London: Macmillan, 1894.

———. *Mendel's Principles of Heredity.* Cambridge: Cambridge University Press, 1902.

———. *Problems of Genetics.* New Haven, Conn.: Yale University Press, 1913.

Berggren, J. W. A., and J. van Couvering, eds. *Catastrophes and Earth History.* Princeton, N.J.: Princeton University Press, 1984.

Bibby, Cyril. "Huxley and the Reception of the 'Origin,' " *Victorian Studies* 3 (1959): 76–86.

———. *T. H. Huxley: Scientist, Humanist and Educator.* New York: Horizon, 1960.

———. *Scientist Extraordinary: The Life and Work of Thomas Henry Huxley.* Oxford: Pergamon Press, 1972.

Bonner, John Tyler, ed. *Evolution and Development.* Berlin: Springer-Verlag. 1982.

Bowler, Peter *Fossils and Progress.* New York: Science History Publications, 1976.

———. "Darwinism and the Argument from Design." *Journal of the History of Biology* 10 (1977): 29–43.

———. *The Eclipse of Darwinism.* Baltimore: Johns Hopkins University Press, 1983.

———. *Evolution, the History of an Idea.* Berkeley: University of California Press, 1984.

———. *The Non-Darwinian Revolution.* Baltimore: Johns Hopkins University Press, 1988.

———. "Development and Adaption: Evolutionary Concepts in British Morphology, 1870–1914." *British Journal for the History of Science* 22 (1989): 283–97.

———. *Life's Splendid Drama.* Chicago: University of Chicago Press, 1996.

Brooke, John H. *Science and Religion: Some Historical Perspectives.* New York: Cambridge University Press, 1991.

Brooks, D., and E. O. Wiley. *Evolution as Entropy.* Chicago: University of Chicago Press, 1986.

Buckland, William. *Reliquiae Diluvianae.* London: John Murray, 1823.

———. *Geology and Mineralogy Considered with Reference to Natural Theology.* Philadelphia: Phillip Carey, Lea and Blanchard, 1837.

Burchfield, Joe. *Lord Kelvin and the Age of the Earth.* Chicago: University of Chicago Press, 1975, 1990.

Campbell, John, and J. William Schopf, eds. *Creative Evolution?!* Boston: Jones and Bartlett Publishers, 1994.

Caneva, Ken. "Teleology with Regret." *Annals of Science* 47 (1990): 291–300.

Cannon, Walter. "The Uniformitarian-Catastrophist Debate."*Isis* 51 (1960): 38–55.

———. "The Impact of Uniformitarianism: Two Letters from John Herschel to Charles Lyell, 1836–1837." *Proceedings of the American Philosophical Society* 105 (1961): 301–14.

———. "Charles Lyell, Radical Actualism, and Theory." *British Journal for the History of Science* 9 (1976): 104–20.

Carpenter, William. *Principles of Physiology*. London: John Churchill, 1853.

———. *Nature and Man*. New York: D. Appleton & Co., 1888.

Chambers, Robert. *Vestiges of the Natural History of Creation*. Surry: Leicester University Press, 1844, 1969.

Charlesworth, Brian, Rusell Lande, and Montgomery Slatkin. "A Neo-Darwinian Commentary on Macroevolution." *Evolution* 36 (1982): 474–98.

Cloyd, E. L. *James Burnett, Lord Monboddo*. Oxford: Clarendon Press, 1972.

Coleman, William. "Lyell and the Reality of Species." *Isis* 53 (1962): 325–38.

———. *George Cuvier, Zoologist*. Cambridge, Mass.: Harvard University Press, 1964.

———. "Morphology Between Type Concept and Descent Theory." *Journal of the History of Medicine* 31 (1976): 149–75.

———. *Biology in the Nineteenth Century: Problems of Form, Function, and Transmutation*. Cambridge: Cambridge University Press, 1977.

Cracraft, Joel, and Niles Eldredge, eds. *Phylogenetic Analysis and Paleontology*. New York: Columbia University Press, 1979.

Crick, Francis. *What Mad Pursuit*. New York: Basic Books, 1988.

Darwin, Charles *The Origin of Species*. London: John Murray, 1859; New York: Avenel, 1976.

———. *The Descent of Man*, 1871; Princeton, N.J.: Princeton University Press, 1981.

———. *The Expression of the Emotions in Man and Animals*, 1872; Chicago: University of Chicago Press, 1965.

———. *The Origin of Species*, 6th ed. London: John Murray, 1872; New York: Collier Books, 1962.

———. *Life and Letters of Charles Darwin*, 2 vols. Francis Darwin, ed. London: John Murray, 1887.

———. *Animals and Plants under Domestication*, 2d ed. New York: D. Appleton & Co., 1892.

————. *More Letters of Charles Darwin*, 2 vols. Francis Darwin and A. C. Seward, eds. London: John Murray, 1903.

————. *Charles Darwin's Autobiography with His Notes and Letters Depicting the Growth of* The Origin of Species. New York: Henry Schuman, 1950.

Dawson, William, ed. *The Huxley Papers*. London: Macmillan & Co., 1946.

de Beer, Galvin. *Homology an Unsolved Problem*. Oxford: Oxford University Press, 1971.

Depew, David, and Bruce Weber. *Darwinism Evolving*. Cambridge, Mass.: MIT Press, 1995.

Desmond, Adrian "Designing the Dinosaur." *Isis* 70 (1979): 224–34.

————. *Archetypes and Ancestors*. Chicago: University of Chicago Press, 1984.

————. *The Politics of Evolution: Morphology, Medicine and Reform in Radical London*. Chicago: University of Chicago Press, 1989.

————. *Huxley: From Devil's Disciple to Evolution's High Priest*. Boston: Addison-Wesley, 1994.

Desmond, Adrian, and James Moore. *Darwin*. London: Michael Joseph, 1991.

Di Gregorio Mario. "Order or Process of Nature: Huxley's and Darwin's Different Approaches to Natural Science." *History and Philosophy of Life Science* 3 (1981): 217–36.

————. "The Dinosaur Connection: A Reinterpretation of T. H. Huxley's Evolutionary View." *Journal of the History of Biology* 15 (1982): 397–418.

————. *T. H. Huxley's Place in Natural Science*. New Haven, Conn.: Yale University Press, 1984.

————. "A Wolf in Sheep's Clothing: Carl Genenbaur, Ernst Haeckel, the Vertebral Theory of the Skull and the Survival of Richard Owen." *Journal of the History of Biology* 28 (1995): 247–80.

Dobzhansky, Theodosius. *Genetics and the Origin of Species*. New York: Columbia University Press, 1937.

Draper, John. *History of the Conflict between Religion and Science*. New York: D. Appleton & Co., 1875.

Ducan, David, ed. *Life and Letters of Herbert Spencer*. 2 vols. London: Methuen and Co., 1908.

Editor's Table. "Professor Huxley's Lectures"; "Professor Huxley on the Horse." *Popular Science Monthly* 10 (1877): 103–104, 369–70.

Eldredge, Niles, and Joel Cracraft. *Phylogenetic Patterns and the Evolutionary Process*. New York: Columbia University Press, 1980.

Eldredge, Niles, and Stephen Gould. "Punctuated Equilibria: An Alternative to Phyletic Gradualism." In *Models in Paleobiology,* T. J. M. Schopf, ed. San Francisco: Freeman, Cooper and Co., 1972.

Ellegard, Alvar. *Darwin and the General Reader.* Goteburg, Germany: Acta Universitatis Gothenburgensis, 1958; Chicago: University of Chicago Press, 1990.

Farber, Paul. "The Type-Concept in Zoology." *Journal of the History of Biology* 9 (1976): 93–119.

Fischman, J. "Feathers Don't Make the Bird." *Discover* (January 1999).

Fitch, Walter. "The Challenges to Darwinism Since the Last Centennial and the Impact of Molecular Studies." *Evolution* 36 (1982): 1133–43.

Flower, William H. "Reminiscences of Professor Huxley." *North American Review* 161 (1895): 279–86.

Forbes, Edward. "On the Manifestation of Polarity in the Distribution of Organic Beings." *Notices of the Proceedings of the Royal Institute,* 1851–1854.

Fortney, R. A. "Gradualism and Punctuated Equilibria as Competing and Complementary Theories." *Special Papers in Paleontology* 33 (1985): 17–28.

Fox-Keller, Elizabeth, and Elizabeth Lloyd, eds., *Key Words in Evolutionary Biology.* Cambridge, Mass.: Harvard University Press, 1992.

Frazetta, T. H. "Hopeful Monsters to Bolyerine Snakes?" *American Naturalist* 104 (1970): 55–72.

Futuyma, Douglas. "Sturm und Drang and the Evolutionary Synthesis." *Evolution* 42, no. 2 (1988): 217–26.

Geison, Gerald. *Michael Foster and the Cambridge School of Physiology.* Princeton, N.J.: Princeton University Press, 1978.

Ghiselin, Michael. *The Triumph of the Darwinian Method.* Berkeley: University of California Press, 1969.

———. "The Individual in the Darwinian Revolution." *New Literary History* 3 (1971): 113–34.

Gilbert, Scott. "Altruism and Other Unnatural Acts: T. H. Huxley on Nature, Man and Society." *Perspectives in Biology and Medicine* 22 (1979): 346–58.

———. "Conceptual Breakthroughs in Developmental Biology." *Journal of the Biological Sciences* 23, no. 3 (1998): 169–76.

Gilbert, Scott, John Opitz, and Rudolf A. Raff. "Resynthesizing Evolutionary and Developmental Biology." *Developmental Biology* 173 (1996): 357–72.

Gillespie, Neal C. *Charles Darwin and the Problem of Creation.* Chicago: University of Chicago Press, 1979.

Gillispie, Charles. *Genesis and Geology: A Study in the the Relations of Scien-

tific Thought, Natural Theology and Social Opinions in Great Britain 1790–1850. New York: Harper, 1959.

Gingerich, Phillip D. "Paleontology and Phylogeny: Patterns of Evolution at the Species Level in Early Tertiary Mammals." *American Journal of Science* 276 (1976): 1–28.

———. "Species in the Fossil Record: Concepts, Trends, and Transitions." *Paleobiology* 11, no. 1 (1985): 27–41.

Glass, Bentley, Otto Temkin, and William Straus Jr., eds. *Forerunners of Darwin: 1745–1859.* Baltimore: Johns Hopkins Press, 1959.

Glick, Thomas, ed. *The Comparative Reception of Darwinism.* Chicago: University of Chicago Press, 1988.

Goldschmidt, Richard. *The Material Basis of Evolution.* New Haven, Conn.: Yale University Press, 1940, 1982.

Goodwin, Brian. *How the Leopard Changed Its Spots.* New York: Charles Schribner's Sons, 1994.

Goodwin, Brian, and Peter Saunders, eds. *Theoretical Biology.* Edinburgh: Edinburgh University Press, 1989.

Gould, Stephen. "Is Uniformitarianism Really Necessary?" *American Journal of Science* 263 (1965): 222–28.

———. "Evolutionary Paleontology and the Science of Form." *Earth Science Review* 6 (1970): 77–119.

———. "The Eternal Metaphors of Paleontology." In *Patterns of Evolution*, A. Hallam, ed., 1–26. Amsterdam: Elsevier, 1977.

———. *Ever Since Darwin.* New York: W. W. Norton and Co., 1977.

———. *Ontogeny and Phylogeny.* Cambridge, Mass.: Harvard University Press, Belknap Press, 1977.

———. "Agassiz's Marginalia in Lyell's *Principles* or the Perils of Uniformity and the Ambiguity of Heros." *Studies in the History of Biology* 3 (1979): 119–38.

———. "Is a New and General Theory of Evolution Emerging?" *Paleobiology* 6 (1980): 119–30.

———. "The Promise of Paleobiology: As a Nomothetic Evolutionary Discipline." *Paleobiology* 6 (1980): 96–118.

———. "But Not Wright Enough: Reply to Orzack." *Paleobiology* 7 (1981): 131–37.

———. "Darwin and the Expansion of Evolutionary Theory." *Science* 216 (1982): 380–87.

———. *The Panda's Thumb.* New York: W. W. Norton and Co., 1982.

———. "The Life and Work of T. J. M. Schopf (1932–1984)." *Paleobiology* 10 (1984): 280–85.

————. *Wonderful Life*. New York: W. W. Norton and Co., 1989.

Gould, Stephen, and Niles Eldredge. "Punctuated Equilibria: The Tempo and Mode of Evolution Reconsidered." *Paleobiology* 3 (1977): 115–51.

Gould, Stephen, and Richard Lewontin. "The Spandrels of San Marco and the Panglossian Paradigm." *Proceedings of the Royal Society of London* 205 (1979): 581–98.

Gould, Stephen, and Elizabeth S. Vrba. "Exaptation—A Missing Term in the Science of Form." *Paleobiology* 8 (1982): 4–15.

Greene, John. *Death of Adam*. Ames: University of Iowa Press, 1959.

————. *Science, Ideology, and World View*. Berkeley: University of California Press, 1981.

Grene, Majorie, ed. *Dimensions of Darwinism*. New York: Cambridge University Press, 1983.

Groeben, Christiane, ed. *Charles Darwin–Anton Dohrn Correspondence*. Naples: Macchiaroli, 1982.

Gruber, Jacob. *A Conscience in Conflict, The Life of St. George Jackson Mivart*. New York: Columbia University Press, 1960.

————. "An Introductory Essay." Owen Manuscripts, British Museum, Natural History, 1960.

Haeckel, Ernst. *Generelle Morphologie der Organism*. 2 vols. Berlin: Reimer, 1866.

————. *The History of Creation*, 8th ed. 2 vols. Translated by Sir Ray Lankester. New York: D. Appleton & Co., 1914.

Hall, Brian K. *Homology*. New York: Academic Press, 1994.

Hallam, Anthony, ed. *Patterns of Evolution as Illustrated by the Fossil Record*. Amsterdam: Elsevier, 1977.

————. "How Rare Is Phyletic Gradualism and What Is Its Evolutionary Significance?" *Paleobiology* 4 (1978): 6–25.

Hilmmelfarb, Gertrude. *Darwin and the Darwinian Revolution*. New York: Doubleday and Co., Inc., 1962.

Hooykaas, Reijer. *Principle of Uniformitarinism in Geology, Biology, and Theology: Natural Law and Divine Miracle*. Leiden: E. J. Brill, 1959.

Hull, David, ed. *Darwin and His Critics*. Cambridge, Mass.: Harvard University Press, 1973.

————. "Darwinism as an Historical Entity: A Historiographical Proposal." *The Darwinian Heritage*, D. Kohn, ed. Princeton, N.J.: Princeton University Press, 1985.

————. *Science as a Process*. Chicago: University of Chicago Press, 1988.

Hutton, James. *Theory of the Earth, Transactions of the Royal Society of Edinburgh* 1 (1788).

Huxley, Julian. *Evolution, the Modern Synthesis*. London: George Allen and Sons, 1942.

Huxley, Thomas. "Vestiges of the Natural History of Creation Tenth Edition London, 1853." *The British and Foreign Medico-Chirurgical Review* 13 (1854): 425–39.

———. "Contemporary Literature: Science." *Westminster Review* 63 (1855): 240.

———. *On Our Knowledge of the Causes of the Phenomena of Organic Nature Being Six Lectures to Working Men*. London: Robert Harwick, 1863.

———. *Lay Sermons, Addresses, and Reviews*. London: Macmillan & Co., 1871.

———. *More Criticisms on Darwin*. New York: D. Appleton & Co., 1872.

———. *Critiques and Addresses*. New York: D. Appleton & Co., 1873.

———. "The Demonstrative Evidence of Evolution." *Popular Science Monthly* 10 (1877): 285–98.

———. "The Hypothesis of Evolution: The Neutral and the Favorable Evidence." In *American Addresses with a Lecture on the Study of Biology*. London: Macmillan & Co., 1877.

———. "The Negative and Positive Evidence." *Popular Science Monthly* 10 (1877): 207–23.

———. "The Three Hypotheses of the History of Nature." *Popular Science Monthly* 10 (1877): 51–56.

———. *Science and Culture*. New York: D. Appleton & Co., 1882.

———. "The Reception of the 'Origin of Species.' " In *Life and Letters of Charles Darwin*. 2 vols. F. Darwin, ed. New York: D. Appleton & Co., 1887.

———. *American Addresses: With a Lecture on the Study of Biology*. New York: D. Appleton & Co., 1888.

———. *Collected Essays of Thomas Huxley*. 9 vols. London: Macmillan & Co., 1893–1898.

———. "Geological Contemporaneity and Persistent Types of Life." In *Discourses Biological and Geological*. London: Macmillan & Co., 1894.

———. "Owen's Position in the History of Anatomical Science." In *The Life of Richard Owen*. 2 vols. Richard Owen, ed. London: John Murrary, 1894.

———. *Scientific Memoirs of Thomas Henry Huxley*. 4 vols. Michael Foster and E. Ray Lancaster, eds. London: Macmillan & Co., 1898–1902.
Volume 1
"Notes on the Medusae Polypes," 1849.

"On the Anatomy and the Affinities of the Family of the Medusae," 1849.
"Zoological Notews and Observations Made on Board HMS *Rattlesnake* During the Years 1846–1850," 1851.
"An Account of Researches into the Anatomy of the Acalephahe," 1851.
"Upon Animal Individuality," 1852.
"Researches into the Structure of the Ascidians," 1852.
"On the Morphology of the Cephalous Mollusca," 1853.
"On the Identity of Structure of Plants and Animals," 1853.
"The Cell Theory," 1853.
"Abstract on the Common Plan of Animal Forms," 1854.
"On Certain Zoological Arguments Commonly Adduced in Favour of the Hypothesis of the Progressive Development of Animal Life in Time," 1854.
"Natural History as Knowledge, Discipline, and Power," 1856.
"On the Theory of the Vertebrate Skull," 1858.

Volume 2

"On the *Stagonolepis Robertsoni* (Agassiz) of the Elgin Sandstones," 1859.
"On the Persistent Types of Animal Life," 1859.
"On the Zoological Relations of Man with the Lower Animals," 1861.
"On New Labyrinthodonts from the Edinburg Coal-Field," 1862.
"Description of *Anthracosaurus Russelli*, a New Labyrinthdont from the Lanarkshire Coal-Field," 1863.

Volume 3

"Explanatory Preface to the Catalogue of the Paleontological Collection in the Museum of Practical Geology," 1865.
"On a New Species of *Telerpeton Elginense*," 1867.
"On the Classification of Birds; and on the Taxonomic Value of the Modifications of Certain of the Cranial Bones Observed in that Class," 1867.
"On the Animals Which are Most Nearly Intermediate between Birds and Reptiles," 1868.
"On *Saurosternon Bainii*, and *Pristerodon McKay*, Two new Fossil Lacertilian Reptiles from South Africa," 1868.
"On the Classification of the Dinosauria with Observations on the Dinosauria of the Trias," 1870.
"On Hyperodapedon," 1869.
"Further Evidence of the Affinity between the Dinosaurian Reptiles and Birds," 1870.

Volume 4

"On the Recent Work of the 'Challenger' Expedition, and Its Bearing on Geological Problems," 1875.
"On the Classification of the Animal Kingdom," 1876.
"On the Evidence as to the Origin of Existing Vertebrate Animals," 1876.
"On the Border Territory between the Animal and the Vegetable Kingdoms," 1876.

———. *Life and Letters of T. H. Huxley.* 2 vols. Leonard Huxley, ed. New York: D. Appleton & Co., 1900.

————. *T. H. Huxley's Diary of the Voyage of HMS* Rattlesnake. Julian Huxley, ed. New York: Doubleday, Doran & Co., 1936.

Irvine, William. *Apes, Angels and Victorians.* New York: Time, 1955.

Jensen, Vernon. *Thomas Henry Huxley: Communicating for Science.* Newark, N.J.: Associated University Press, 1991.

Jepson, Glenn, Ernst Mayr, and George Gaylord Simpson, eds. *Genetics, Paleontology and Evolution.* Princeton, N.J.: Princeton University Press, 1949.

Kauffman, Stuart. "Origins of Order in Evolution, Self-Organisation, and Selection." In *Theoretical Biology,* Brian Goodwin and Peter Saunders, eds. Edinburgh: Edinburgh University Press, 1989.

————. *At Home in the Universe.* New York: Oxford University Press, 1995.

Kimura, Motoo. *The Neutral Theory of Molecular Evolution.* Cambridge: Cambridge University Press, 1983.

Kingsley, Charles. *The Waterbabies.* 1872; London: J. M. Dent & Sons, 1985.

Kitts, David. "Paleontology and Evolutionary Theory." *Evolution* 28 (1974): 458–72.

Kohn, David, ed. *The Darwinian Heritage.* Princeton, N.J.: Princeton University Press, 1985.

Kottler, Malcolm. "Alfred Russel Wallace, the Origin of Man, and Spiritualism." *Isis* 65 (1974): 145–92.

————. "Charles Darwin and Alfred Russel Wallace: Two Decades of Debate over Natural Selection." In *The Darwinian Heritage,* D. Kohn, ed. (Princeton, N.J.: Princeton University Press, 1982), pp. 365–432.

Lamarck, Jean Baptiste. *Zoological Philosophy.* 1809; Chicago: University of Chicago Press, 1984.

Lande, Russell. "Expected Time for Random Genetic Drift of a Population between Stable Phenotypic States." *Proceedings of the National Academy of Science* 82 (1985): 7641–45.

————. "The Dynamics of Peak Shifts and the Pattern of Morphological Evolution." *Paleobiology* 12, no. 4 (1986): 343–54.

Lenoir, Timothy. *The Strategy of Life.* Dordrecht, The Netherlands: D. Reidel Publishing Co., 1982.

Levinton, Jeffrey. "Stasis in Progress: The Empirical Basis of Macroevolution." *Annual Review of Ecological Systems* 14 (1981): 113.

Levinton, Jeffrey, and Douglas Futuyma. "Macroevolution in Pattern and Process, Introduction and Background." *Evolution* 36, no. 3 (1982): 425–26.

Levinton, Jeffrey, and Chris Simon. "A Critique of the Punctuated Equilibria Model and Implications for the Detection of Speciation in the Fossil Record." *Systematic Zoology* 29, no. 2 (1980): 130–42.

Levinton, Jeffrey, et al. "Organismic Evolution: The Interaction of Micro-evolutionary and Macroevolutionary Processes." In *Patterns and Processes in the History of Life*, David Raup and David Jablonski, eds. Berlin: Springer-Verlag, 1986.

Lewontin, Richard, Steven Rose, and Leon Kamin. *Not in Our Genes.* New York: Pantheon, 1984.

Lightman, Bernard. *The Origins of Agnosticism: Victorian Unbelief and the Limits of Knowledge.* Baltimore: Johns Hopkins University Press, 1987.

———, ed. *Victorian Science in Context.* Chicago: University of Chicago Press, 1997.

Lindberg, David, and Ronald Numbers, eds. *God and Nature.* Berkeley: University of California Press, 1986.

Lyell, Charles. *Principles of Geology*, vols. 1–3. London: John Murray, 1830–1834; Chicago: University of Chicago Press, 1990.

———. "Anniversary Address of the President." *Proceedings of the Geological Society of London* 7 (1851): xxv–lxxvi.

———. *Geological Evidences of the Antiquity of Man.* Philadelphia: George W. Childs, 1863.

Lyell, Katherine, ed. *Sir Charles Lyell: Life, Letters and Journals.* 2 vols. London: John Murray, 1881.

Lyons, Sherrie. "Thomas Huxley: Fossils, Persistence, and the Argument from Design." *Journal of the History of Biology* 26 (1993): 545–69.

———. "The Origins of T. H. Huxley's Saltationalism: History in Darwin's Shadow." *Journal of the History of Biology* 28 (1995): 463–94.

———. Review of Nicolaas Rupke's *Victorian Naturalist. Quarterly Review Biology* 70 (1995): 326–27.

———. "Convincing Men They Are Monkeys." In *Thomas Henry Huxley's Place in Science and Letters*, Alan Barr, ed. Athens: University of Georgia Press, 1997.

———. "Taxonomy Recapitulates Society." (Review of Harriet Ritvo's *The Platypus and the Mermaid and Other Figments of the Classifying Imagination*.) *Science* 279 (January 2, 1998): 38.

Maderson, P. F. A., et al., "The Role of Development in Macroevolutionary Change." In *Evolution and Development*, John Tyler Bonner, ed. New York: Springer-Verlag, 1982.

Maienschein, Jane. *Transforming Traditions in American Biology, 1880–1915.* Baltimore: Johns Hopkins University Press, 1991.

Marshall, Charles, and J. William Schopf, eds. *Evolution and the Molecular Revolution.* Boston: Jones and Bartlett Publishers, 1996.

Mayr, Ernst. *Systematics and the Origin of Species.* New York: Columbia University Press, 1942.

———. *Animal Species and Evolution.* Cambridge, Mass.: Harvard University Press, 1963.

———. *Evolution and the Diversity of Life.* Cambridge, Mass.: Harvard University Press, 1976.

———. *The Growth of Biological Thought: Diversity, Evolution and Inheritance.* Cambridge, Mass.: Harvard University Press, 1982.

———. "Speciation and Macroevolution." *Evolution* 36 (1982): 1119–32.

———. "Darwin's Five Theories of Evolution," in *The Darwinian Heritage,* D. Kohn, ed. Princeton, N.J.: Princeton University Press, 1985.

———. *Towards a New Philosophy of Biology.* Cambridge, Mass.: Harvard University Press, 1988.

———, ed. *The Species Problem: Symposium of the AAAS, 1955.* Washington, D.C.: AAAS Publication #50, 1957.

Mayr, Ernst, and William Provine, eds. *The Evolutionary Synthesis.* Cambridge, Mass.: Harvard University Press, 1980.

McKinney, H. Lewis. "Alfred Russel Wallace and the Discovery of Natural Selection." *Journal of the History of Medicine* 21 (1966): 333–57.

———. *Wallace and Natural Selection.* New Haven, Conn.: Yale University Press, 1972.

Milne, D., D. Raup, J. Billingham, K. Nillaus, and K. Padian, eds. *The Evolution of Complex and Higher Organisms.* Washington, D.C.: NASA, 1985.

Miramontes, O., R. Sole, and B. Goodwin. "Collective Behavior of Random-Activated Mobile Cellular Automata." *Physica* 63 (1993): 145–60.

Mivart, George J. "Darwin's Descent of Man." *Quarterly Review* 131 (July 1871): 47–90. Also published in David Hull, ed., *Darwin and His Critics.* Cambridge, Mass.: Harvard University Press, 1973.

———. *On the Genesis of Species.* London: Macmillan, 1871.

———. "Evolution in Professor Huxley." *Nineteenth Century* 34 (1893): 198–211.

Moore, James. *The Post-Darwinian Controversies.* Cambridge: Cambridge University Press, 1979.

———, ed. *History, Humanity and Evolution: Essays in Honor of John Greene.* Cambridge: Cambridge University Press, 1989.

Nicolis, G., and I. Prigone. *An Introduction to Complexity.* New York: W. H. Freeman, 1987.

Nitecki, Matthew, ed. *Evolutionary Progress.* Chicago: University of Chicago Press, 1989.

————. *Evolutionary Innovations.* Chicago: University of Chicago Press, 1990.

Nyhart, Lynn. *Biology Takes Form: Animal Morphology and the German Universities 1800–1900.* Chicago: University of Chicago Press, 1995.

Oldroyd David, and I. Langham, eds. *The Wider Domain of Evolutionary Thought.* Dordrecht, The Netherlands: D. Reidel Publishing Co., 1983.

Olson, Evertt. "The Problem of Missing Links: Today and Yesterday." *Quarterly Review of Biology* 56 (1981): 405–38.

Oppenheimer, Jane. *Essays in the History of Embryology and Biology.* Cambridge, Mass.: MIT Press, 1967.

Orzack, Steven. "The Modern Synthesis Is Partly Wright." *Paleobiology* 7 (1981): 128–34.

Ospovat, Dov. "The Influence of Karl Ernst von Baer's Embryology, 1828–859: A Reappraisal in Light of Richard Owen and William B. Carpenter's Paleontological Application of von Baer's Law." *Journal of the History of Biology* 9 (1976): 1–28.

————. "Perfect Adaptation and Teleological Explanation." *Studies in the History of Biology* 2 (1978): 33–56.

————. *The Development of Darwin's Theory: Natural History, Natural Theology, and Natural Selection, 1838–1839.* Cambridge: Cambridge University Press, 1981.

Oster, G., and P. Alberch. "Evolution and Bifurcation of Developmental Programs." *Evolution* 36, no. 3 (1982): 444–59.

Owen, Richard. "Archetype and Homologies of the Vertebrate Skeleton." *British Association for the Advancement of Science Report,* 1846.

————. *On the Nature of Limbs.* London: n.p., 1849.

————. "Lyell—On Life and Its Successive Development." *Quarterly Review* 89 (1851): 412–51.

————. *Principal Forms of the Skeleton and of the Teeth.* Philadelphia: Blanchard and Lea, 1854.

————. "On the Characters, Principles of Division, and Primary Groups of the Class Mammalia." *Journal of the Proceedings of the Linnean Society* (Zoo.) 2 (1858): 1–37.

————. "On the Fossil Remains of a Longtailed Bird (*Archaeopteryx macrurus*) from the Lithographic Slate of Solenhofen." *Proceedings of the Royal Society of London* 12 (1862): 272–74.

Owen, Rev. Richard, ed. *The Life of Richard Owen.* 2 vols. London: John Murray, 1894.

Paradis, James. *T. H. Huxley: Man's Place in Nature.* Lincoln: University of Nebraska Press, 1978.

Paradis, James, and G. C. Williams, eds. *T. H. Huxley's "Evolution and Ethics": With New Essays on Its Victorian and Sociobiological Context.* Princeton, N.J.: Princeton University Press, 1989.

Patterson, Colin. *Molecules and Morphology in Evolution.* Cambridge: Cambridge University Press, 1987.

Pingree, J. *Thomas Henry Huxley, A List of His Scientific Notebooks, Drawings, and Other Papers.* London: Imperial College, 1968.

Provine, William. *The Origins of Theoretical Population Genetics.* Chicago: University of Chicago Press, 1971.

———. *Sewall Wright: Geneticist and Evolutionist.* Chicago: University of Chicago Press, 1986.

———. "Progress in Evolution and Meaning in Life." *Evolutionary Progress,* M. Nitecki, ed., pp. 49–74. Chicago: University of Chicago Press, 1989.

[Pycroft, George]. "A Sad Case, Mansion House." Owen Manuscripts, British Museum of Natural History, April 23, 1863.

Raff, Rudolf. *The Shape of Life.* Chicago: University of Chicago Press, 1996.

Raff, Rudolf A., and E. C. Raff, eds. *Development as an Evolutionary Process.* New York: Alan Liss, 1987.

Raup, David. "On the Early Origins of Major Biologic Groups." *Paleobiology* 9, no. 2 (1983): 107–15.

———. *The Nemesis Affair.* New York: W. W. Norton, 1986.

Raup, David, and Stephen J. Gould. "Stochastic Simulation and Evolution of Morphology Towards a Nonmothetic Paleontology." *Systematic Zoology* 23 (1974): 305–22.

Raup, David, and David Jablonski, eds. *Patterns and Processes in the History of Life.* Dahlem Konferenzen, Berlin: Springer-Verlag, 1986.

Rehbock, Philip. "Huxley, Haeckel, and the Oceanographers: The Case of *Bathybius haeckeli.*" *Isis* 66 (1975): 504–33.

———. *The Philosophical Naturalists.* Madison: University of Wisconsin Press, 1983.

Rensch, Bernhard. *Evolution Above the Species Level.* New York: Columbia University Press, 1959.

Rhodes, Frank. "Gradualism, Punctuated Equilibria and *The Origin of Species.*" *Nature* 305 (1983): 269–72.

———. "Gradualism and Its Limits: The Development of Darwin's Views on the Rate and Pattern of Evolutionary Change." *Journal of the History of Biology* 202 (1987): 139–57.

Richards, Evelleen. "A Question of Property Rights: Richard Owen's Evolutionism Reassessed." *British Journal for the History of Science* 20 (1987): 129–71.

————. "Huxley and Woman's Place in Science." In *History, Humanity and Evolution*, J. Moore, ed. Cambridge: Cambridge University Press, 1989.

Richards, Robert. *Darwin and the Emergence of of Evolutionary Theories of Mind and Behavior*. Chicago: University of Chicago Press, 1987.

————. *The Meaning of Evolution*. Chicago: University of Chicago Press, 1992.

Ritvo, Harriet. *The Platypus and the Mermaid and Other Figments of the Classifying Imagination*. Cambridge, Mass.: Harvard University Press, 1997.

Roller, Duane, ed. *Perspectives in the History of Science and Technology*. Norman: University of Oklahoma Press, 1971.

Romanes, George. "Physiological Selection: An Additional Suggestion on the Origin of Species." *Journal of the Linnean Society of London* (Zoo.) 19 (1886): 337–411.

Roth, Louise. "Homology and Hierarchies: Problems Solved and Unresolved." *Journal of Evolutionary Biology* 4 (1991): 167–94.

Rudwick, Martin. "Uniformity and Progression: Reflections on the Structure of Geological Theory in the Age of Lyell." In *Perspectives in the History of Science and Technology*, Duane H. Roller, ed., 209–27. Norman: University of Oklahoma Press, 1971.

————. *The Meaning of Fossils*. Chicago: University of Chicago Press, 1982.

————. *The Great Devonian Controversy*. Chicago: University of Chicago Press, 1985.

————. *Scenes from Deep Time*. Chicago: University of Chicago Press, 1992.

Rupke, Nicolaas. "Richard Owen's Vertebrate Archetype." *Isis* 84 (1993): 231–51.

————. *Richard Owen Victorian Naturalist*. New Haven, Conn.: Yale University Press, 1994.

Ruse, Michael. *The Darwinian Revolution*. Chicago: University of Chicago Press, 1979, 1981.

————. *From Monad to Man*. Cambridge, Mass.: Harvard University Press, 1997.

Russell, E. S. *Form and Function: A Contribution to the History of Animal Morphology*. 1916; Chicago: University of Chicago Press, 1982.

Salthe, Stanley. *Development in Evolution: Complexity and Change in Biology*. Cambridge, Mass.: MIT Press, 1993.

Scheibel, Arnold, and J. William Schopf, eds. *The Origin and the Evolution of Intelligence*. Boston: Jones and Bartlett, 1997.

Schopf, Thomas J. M., ed. *Models in Paleobiology*. San Francisco: Freeman, Cooper and Co., 1972.

———. "Punctuated Equilibria and Evolutionary Stasis." *Paleobiology* 7 (1981): 156–66.

———. "A Critical Assessment of Punctuated Equilibria I: Duration of Taxa." *Evolution* 36 (1982): 1133–54.

Secord, James. *Controversies in Victorian Geology.* Princeton, N.J.: Princeton University Press, 1986.

Seilacher, A. "Early Multicellular Life, Late Proterozoic Fossils and the Cambrian Explosion." In *Early Life on Earth*, S. Bengston, ed., pp. 389–400. New York: Columbia University Press, 1994.

Simpson, George Gaylord. *Tempo and Mode in Evolution.* New York: Columbia University Press, 1944.

———. *The Major Features of Evolution.* New York: Columbia University Press, 1953.

Skarda, C. A., and W. F. Freeman. "How Brains Make Chaos in Order to Make Sense of the World." *Behavior and Brain Science* 10 (1989): 161–95.

Sloan, Phillip. "Darwin, Vital Matter, and the Transformism of Species." *Journal of the History of Biology* 19 (1986): 369–445.

———, ed. *Richard Owen: The Hunterian Lectures in Comparative Anatomy May–June, 1837, With an Introductory Essay and Commentary.* Chicago: University of Chicago Press, 1992.

Sober, Elliot. *Conceptual Issues in Evolutionary Biology.* Cambridge, Mass.: MIT Press, 1983.

———. *The Nature of Selection.* Cambridge, Mass.: MIT Press, 1985.

Somit, Albert, and Stevenson Peterson, eds. *The Dynamics of Evolution: The Punctuated Equilibrium Debate in the Natural and Social Sciences.* Ithaca, N.Y.: Cornell University Press, 1992.

Spencer, Herbert. "The Proper Sphere of Government." *Nonconformist* (June 1842–November 1842).

———. *Social Statics.* 1850; New York: Robert Schalkenbach Foundation, 1954.

———. *Principles of Biology.* 1899; Osnabruck: Proff & Co., 1966.

———. *An Autobiography.* 2 vols. New York: D. Appleton & Co., 1904.

Stanley, Steven M. "A Theory of Evolution above the Species Level." *Proceedings of the National Academey of Science* 72 (1975): 646–50.

———. "Chronospecies's Longevities, the Origin of Genera, and the Punctuational Model of Evolution." *Paleobiology* 4 (1978): 26–40.

——— *Macroevolution: Pattern and Process.* San Fransisco: W. H. Freeman & Co., 1979.

———. *The New Evolutionary Timetable.* New York: Basic Books Inc., 1981.

————. "Macroevolution and the Fossil Record." *Evolution* 36 (1982): 460–73.

————. "Rates of Evolution." *Paleobiology* 11, no. 1 (1985): 13–26.

Stebbins, G. Ledyard. "Perspectives in Evolutionary Theory." *Evolution* 36 (1982): 1109–18.

Stebbins, G. Ledyard, and Franciso Ayala. "Is a New Evolutionary Synthesis Necessary?" *Science* 213 (1981): 967–71.

Stennis, C. G. J. van. "Plant Speciation in Milesia, with Special Reference to the Theory of Non-Adaptive Saltatory Evolution." *Biology Journal of the Linnean Society* 1 (1969): 97–133.

Stocking, George. *Race, Culture and Evolution*. Chicago: University of Chicago Press, 1982.

————. *Victorian Anthropology*. New York: Free Press, 1987.

Swenson, Ron. "Thermodynamics and Evolution." In *The Encyclopedia of Comparative Psychology*, G. Greenberg and M. Haraway, eds. New York: Garland Publishing, 1997.

Swetlitz, Marc. "Julian Huxley and the End of Evolution." *Journal of the History of Biology* 28 (1995): 181–217.

Symondson, Anthony, ed. *The Victorian Crisis in Faith: Six Lectures by Robert M. Young and Others*. London: The Society for Promoting Christian Thought, 1970.

Tax, Sol, ed. *Evolution after Darwin*, vols. 1–3. Chicago: University of Chicago Press, 1960.

Thompson, D'Arcy. *On Growth and Form*. John Bonner, ed., abridged edition. Cambridge: Cambridge University Press, 1961.

Turner, Frank. *Between Science and Religion*. New Haven, Conn.: Yale University Press.

Valentine, J. "Fossil Record of the Origin of Bauplane and Its Implications." In *Patterns and Processes in the History of Life*, D. Raup and D. Jablonski, eds. Dahlem Konferenzen, Berlin: Springer-Verlug, 1986.

Vrba, Elizabeth, and Stephen Gould. "The Hierarchical Expansion of Sorting and Selection: Sorting and Selection Cannot Be Equated." *Paleobiology* 12, no. 2 (1986): 217–28.

Wake, David. "Comparative Terminlogy." *Science* 265 (1994): 268–69.

Wallace, Alfred Russel. *Darwin–Wallace Celebration Held on Thursday, 1st July, 1908, by the Linnean Society of London*. London: The Linnean Society, 1864.

————. "The Origin of the Human Races and the Antiquity of Man Deduced from the Theory of Natural Selection." *Journal of the Anthropological Society of London* 2 (1864): clvii-clxxxvii.

———. "Sir Charles Lyell on Geological Climates and the Origin of Species." *Quarterly Review* 126 (1869): 359–94.

———. *Darwinisim: An Exposition of the Theory of Natural Selection with Some of Its Applications.* London: Macmillan, 1889.

Webster, Gerry, and Brian Goodwin. *Form and Transformation.* Cambridge: Cambridge University Press, 1996.

Whewell, William. "Principles of Geology, vol. ii (1832)." *Quarterly Review* 47 (1832): 117.

White, Andrew. *A History of the Warfare of Science with Theology in Christendom.* New York: Appleton, 1896; Amherst, N.Y.: Prometheus Books, 1993.

Williamson, P. G. "Paleontological Documentation of Speciation in Cenozoic Mollusks from Turkana Basin." *Nature* 293 (1981): 437–43.

Wilson, Leonard. *Charles Lyell, the Years to 1841.* New Haven, Conn.: Yale University Press, 1972.

———, ed. *Sir Charles Lyell's Scientific Journals on the Species Question.* New Haven, Conn.: Yale University Press, 1970.

Winfree, A. T. *The Geometry of Biological Time.* New York: Springer-Verlag, 1989.

Winsor, Mary. *Starfish, Jellyfish, and the Order of Life.* New Haven, Conn.: Yale University Press, 1976.

———. "The Impact of Darwinism upon the Linnaean Enterprise, with Special Reference to the Work of T. H. Huxley." In *Contemporary Perspectives on Linnaeus*, John Weinstock, ed. Lanham, Maryland: University Press of America, 1985.

Wright, Sewall. "Genic and Organismic Selection." *Evolution* 34 (1980): 825–43.

———. "Character Change, Speciation, and the Higher Taxa." *Evolution* 36, no. 3 (1982): 427–43.

Young, Bruce. "On the Necessity of an Archetypal Concept in Morphology: With Special Reference to the Concepts of 'Structure' and 'Homology.' " *Biology and Philosophy* 8 (1993): 225–48.

Index

W

Wallace, Alfred Russel, 99, 136, 246, 248, 268, 273, 282

Ward, W. G., 32

Waterbabies, The, 217–19, 223

Whewell, William, 51–52, 99, 241

White, Andrew, 202

Whitworth gun, 43

Wilberforce, Bishop Samuel, 31, 201–202, 210, 219, 221–22, 267

Winsor, Mary, 72–73

Wollaston, Thomas Vernon, 102

X

X Club, 31–32